MATH and *ART*

AN INTRODUCTION TO
VISUAL MATHEMATICS

MATH and *ART*

An Introduction to
Visual Mathematics

Sasho Kalajdzievski

University of Manitoba
Winnepeg, Canada

In collaboration with
R. Padmanabhan

CRC Press
Taylor & Francis Group
Boca Raton   London   New York

CRC Press is an imprint of the
Taylor & Francis Group, an **informa** business

A CHAPMAN & HALL BOOK

Cover image from Jos Leys, *Indra* 461, 2005.

Chapman & Hall/CRC
Taylor & Francis Group
6000 Broken Sound Parkway NW, Suite 300
Boca Raton, FL 33487-2742

© 2008 by Taylor & Francis Group, LLC
Chapman & Hall/CRC is an imprint of Taylor & Francis Group, an Informa business

International Standard Book Number-13: 978-1-58488-913-7 (Softcover)

Library of Congress Cataloging-in-Publication Data

Kalajdzievski, Sasho.
   Math and art : an introduction to visual mathematics / Sasho Kalajdzievski.
      p. cm.
   Includes bibliographical references and index.
   ISBN 978-1-58488-913-7 (alk. paper)
   1. Geometry. 2. Design. 3. Geometry in art. I. Title.

   QA445.K32 2008
   510--dc22   2007046285

Visit the Taylor & Francis Web site at
http://www.taylorandfrancis.com

and the CRC Press Web site at
http://www.crcpress.com

Dedication

To Nina, and to our kids Darja, Damjan, and Timjan. With love.

Во сломен на моите родители, Илинка и Мате.

Contents

Introduction

THE SOURCE

In the winter of 2003, I started teaching a rather unusual course. The course was called Math in Art, and it consisted mainly of visually interesting mathematical topics. The organization of the course was unorthodox: one-third of the course was delivered by an artist or an architect, two-thirds of it was mathematical and was done by a mathematician. The mathematician would present some basic mathematical concepts, and the artist or architect would follow up with some rather loose artistic interpretations of these concepts. It was a relatively new course at that time, introduced only a couple of years earlier. Initially, the course was geared toward freshman with artistic ambitions, but by the time I started teaching it, about one-half of the enrollment included students registered throughout the faculty of science.

The course was designed by a committee of both mathematicians and artists, but the mathematical content of the course was primarily a creation of Dr. R. Padmanabhan. The first student-notes were also compiled by Dr. Padmanabhan. In the summer of 2003, I wrote a new set of notes for the course. The book in your hands is a four-year outgrowth of these notes.

The skeleton of these notes, as well as of the subsequent expanded versions that include this book, is the original content of the Math in Art course. It follows as a corollary that the basic idea for this book belongs to Dr. Padmanabhan. He was always very generous with his encouragement and moral support during the long months of writing, rewriting, and expanding the notes. We often talked about the subjects in the notes, and his advice and insights were valuable to me. He also contributed a number of exercises, graphics, and constructions.

GOAL AND TARGET

The primary goal of this book is to indicate the potential of mathematics for generating (visually) appealing objects, thus providing themes for artistic explorations. The second, equally important goal is to reveal some of the visual beauty of mathematics. The prominence of visual beauty of mathematics, as a criterion for choosing the topics we include here, defines and justifies the word "visual" in the title of this book.

Unequivocally, this is a book about mathematics, not about art. Artwork that we present here is mostly to illustrate the potential of mathematics for artistic inspiration. The graphics and images by various artists included in this book by no means constitute an attempt to categorize mathematically inspired art, neither is it a comprehensive review of such artwork. On the other hand, even though this is primarily a book about mathematics, our goal is not necessarily only to teach it, but rather to convey that mathematics is worth studying. Indeed, a number of times we will encounter mathematical subjects, whose rigorous study would take us way beyond the reach of elementary mathematics; in such cases, we will only provide outlines of the concept and directions to go if one wants to continue the exploration.

Our target audience consists of students of art or mathematics, and, more generally, any one who takes an interest in these two subjects.

To art students we offer a few sources of inspiration and creative ideas. Understanding the mathematical background of some relatively complicated but interesting visual objects (as are, say, fractals) may be an important component in their development. Our experience tells us that art-oriented students find the content of the notes (and the associated course) to be a relevant part of their studies. This is especially highlighted today with the emergence and the growth of computer-generated graphics as a way of expressing artistic ideas.

The message that we try to convey to students of mathematics is that mathematics is literally beautiful. Our goal is to give a gentle, visual introduction to a selection of mathematical themes, to indicate along the way various directions for further exploration. For example, very few students have even cursory encounters with such a central mathematical discipline as topology. We regard these notes (and the course that we teach) as an opportunity to advertise various elegant mathematical theories and notions. Moreover, we also make a tentative effort to dispel the folk myth of equating math with basic arithmetic. Indeed, there are very few numerocentric concepts in this book, and in the rare instances where numbers come to us naturally (as is the case with the golden mean), we put the stress on construction and visualization, almost disregarding all computational aspects.

CONTENT

This book evolved from course notes, and that—being course notes—is one of its primary goals. However, I have tried to make it as accessible as possible, so that almost all of the material is self-sufficient and could easily be used as independent reading material. Even the so-called *mathematical sections* are done in such a way that almost all of the material there does not depend on anything other than *common sense*.

The topics are chosen on the basis of accessibility, their visualization potential, and the mathematical interconnectedness between them. The dependence between the chapters and sections is occasionally rather weak. For example, the last chapter (Topology) is virtually independent from the previous chapters. However, the chapters are ordered roughly according to the level of basic mathematical maturity needed to appreciate the material covered.

We include three *mathematical* sections (2.2, 3.2, and 5.2), where we introduce more abstract structures (complex numbers, matrices). We prominently indicate in the titles of these sections that they are *optional*, meaning that a reader who has difficulty with relatively abstract mathematical ideas could safely skip them without fatally affecting the understanding of the rest of the material. Our goal in these sections is to indicate to the reader how certain basic theories could be used to further investigate the concepts we have informally introduced in the rest of the book. A course that would cover some of these mathematical sections would obviously be more ambitious than the minimal course we describe below.

Here is a brief summary of the Math in Art course taught at the University of Manitoba. The course lasts one semester, but, as indicated above, only two-thirds of the course lectures are devoted to mathematical content (about twenty-one 50 min classes). In this course, we cover the following topics:

1. Basics of Euclidean geometry, golden ratio, Fibonacci numbers (Chapter 1). We pay attention to the last, fifth axiom of Euclidean planar geometry, but we do not develop the theory formally.

2. Symmetries (Chapter 2, Sections 2.1 and 2.3). The stress here is on the visual constructions rather than on formally developing the theory. As a consequence, we give very few proofs. For example, most of the time we simply look for and exhibit groups of symmetries of planar objects, without formally showing that these groups of symmetries exhaust all of the symmetries of the planar objects.

3. Frieze patterns and tilings of the plane (Chapter 2, Sections 2.4 and 2.5). We sometimes cover the groups of symmetries of some simple frieze patterns (and include such questions in the tests). That is as far as we dare to go, given the amount of time we have for mathematics in the course. We talk about tilings, but rarely test students on this topic. However, we do deliver a simple proof in this part of the course, showing that there are only three regular tilings.

4. Similarities (Chapter 3, Section 3.1). This is sometimes covered briefly, depending on the instructor.

5. Fractals (Chapter 3, Sections 3.3 and 3.4). We give examples of tree-type fractals and Koch-type fractals. Occasionally, some of instructors cover problems combining similarities with fractals. Julia sets and complicated fractals are rarely covered—again depending on time and the instructor's discretion.

6. "Hyperbolic geometry" (Chapter 4, Sections 4.1, 4.3, and 4.4). The axioms are covered briefly, putting emphasis on the fifth axiom. We do not cover inversion (Section 4.2) separately; the construction of points under circular inversions is simply embedded within the construction of hyperbolic lines. Most of the time devoted to this section is used for various constructions involving hyperbolic lines. The claim that the sum of the interior angles in hyperbolic triangles is less than 180° is used as a motivation to show some pictures of hyperbolic tilings.

7. Perspective drawing (Chapter 5, Section 5.1). We give the basic rules and do a few examples.

8. Polyhedra and platonic solids (Chapter 5, Section 5.3). Here, we show illustrations or models of various polyhedra. We usually outline a justification of the existence of only five platonic solids. We introduce the Euler characteristic at this point.

9. Conic sections (Chapter 5, Section 5.4). We define (visually) the conic sections, and then we show how to construct (with a ruler and a compass) their approximations.

10. "Topology" (Chapter 6). This is mostly done at the level of show-and-tell. A typical test exercise from this part of the course would require from the student to produce a few intermediate steps illustrating that two objects are homotopic (in the plane or in three dimensions).

The curriculum of the Math in Art course outlined above is by no means superior to other combinations of topics. It is given here only as reference, and it could be considered as a *minimal curriculum* for a one-semester course on the subject. A more ambitious course could (and should) involve more subjects, should go more in depth, and could be done more formally. For example, we have never covered cellular automata, have covered fractals very informally, have not even mentioned inversions explicitly, and have never seriously dwelled on the subject of topology (the level of the first three exercises at the end of Section 6.1 is as deep as we have ever gone in testing that part of the material in our course). The *mathematical* sections (Sections 2.2, 3.2, and 5.2, indicated as optional in this book) are also beyond the scope of the course we currently teach.

REGARDING WRITING THIS BOOK

Even the word "writing" in the above heading is misleading: it may be more fitting to say that this is more a book of illustrations with some words of explanation. It was a challenge to "write" the many illustrations. It often took many days to create a program to generate a visual output, and in a number of cases the result of that effort accounted for one or zero number of illustrations in this book. For example, it took me weeks to write a program for generating a few types of Ammann aperiodic tilings, and the result of that work is one inconspicuous illustration that appears at the end of Section 2.5. I am by no means a programmer or a graphics designer, and I do not feel guilty for *wasting* hours upon hours for, say, a programming endeavor that perhaps could have been done in minutes by an apt professional. On the other hand, it was awarding to see beautiful objects emerging on the computer screen following long, winding work, even in cases when these pictures did not find their way into this book.

ORGANIZATION AND NOTATION

Virtually all of the material in the book is self-contained. On occasion, we do assume knowledge of some of high-school mathematics, but we do not shy from going into areas of elementary high-school mathematics, and reviewing parts of it.

The book contains a number of mathematical overtures, consisting of simple proofs or short tidbits, sparsely distributed throughout the book (we start them with the phrase "A bit of math").

There are many illustrations in this book. They are labeled, and often contain captions. Illustrations within proofs or constructions are called "Steps," whereas more independent illustrations are labeled as "Figures." Since this is a book about visual mathematics, we often prefer pictures to verbal explanations. The electronic version contains a few animations (indicated in the captions of the associated illustrations in the book). There is, obviously, a great potential in going in the direction of animating various claims, proofs, and notions in mathematics.

In order to increase the clarity of the exposition, we indicate the ends of short arguments or examples with a square ($\square$).

Most of the sections end with exercises, and most of these exercises are rather elementary. But there are a number of not-so-easy problems, which we distinguish with asterisks (as in 9*). The few exercises that depend on the mathematical sections (those indicated as being *optional*), appearing out of the mathematical sections, are clearly identified as such.

Occasionally, it was hard to resist the temptation not to cover the mathematical material that was almost within arm-reach. For example, we do talk rather extensively (Chapter 2) about groups of symmetries, but we never define the notion of a *group* as an algebraic structure. A few times, this conscious obstinacy from talking *real* mathematics was the cause of substantial discomfort. It is virtually impossible, to note a prime example, to make the notion of *compactness* (appearing in a footnote in Section 6.2) accessible without losing much of the rigor. In such cases, we sometimes indicated in footnotes the proper mathematical setting and the area of mathematics that needs to be studied.

I have decided not to give references to books, a few exceptions notwithstanding. It would be misleading to quote a mathematical book where some of the subjects that we have covered here are expanded upon, because such books require more serious mathematical background than what is given in this book. So, we will be satisfied with only pointing the direction of further studies, and by identifying branches of mathematics where certain concept that we encounter are studied more formally.

ACKNOWLEDGMENTS AND CREDITS

I thank the following people who read the notes in various stages of completion, and whose suggestions, corrections, and advices were invaluable to me: Dr. R. Padmanabhan, Dr. David Gunderson, Dr. Michelle Davidson, Dr. Peter Penner, Dr. Mile Krajcevski, and Dr. Nina Zorboska. Dr. Padmanabhan's contributions have already been mentioned in the beginning of this introduction. Dr. Penner graciously reviewed the notes several times, as did Dr. Gunderson, who also provided me with a number of books and references. I am also grateful to Dr. Krajcevski for his insightful commentaries and advices. Finally, Dr. Zorboska's detailed commentaries were of vital importance to me.

I am especially thankful to Darja Kalajdzievska for her very detailed and thorough reviews of the notes.

My thanks to David Lucas and Treble Lysenko who have occasionally directed me to interesting artworks.

The images from *Art Explosion 525,000* were used in Figures 1.4.2 through 1.4.5, 2.1.8 through 2.1.11, and 2.5.4 through 2.5.6.

I am grateful to the following artists who gave me their generous permissions to include their artwork. (The copyright owners are the authors unless otherwise stated.)

- Jos Leys, Figures 4.2.13, 4.5.11, 5.5.3, 6.3.9, and 6.3.14. http://www.josleys.com
- Dick Termes, Figures 5.2.12 and 5.4.4. http://www.termespheres.com
- Jos de Mey, Figure 5.1.24
- John Osborn, Figure 2.6.10. http://www.ozbird.net/
- Charles O. Perry, Figures 5.4.18 and 6.3.24. http://www.charlesperry.com/
- George W. Hart, Figures 5.3.12 and 5.3.13. http://www.georgehart.com
- Marcus Vogt, Figures 2.5.7 and 5.5.12. http://www.markusvogt.eu/
- James J. Lemon, Figure 2.6.17. http://www.jjlg.com/
- Jean-Charles Marteau, Figure 5.5.10
- Arend Nijdam, Figure 3.2.6
- Marc Thomson, Figures 3.1.7, 3.3.13, and 3.5.4. www.Marcs-ArtworkStudio.com
- Ken Knowlton, Figure 2.5.8. This artwork is tribute and gift to owner/president of Malden Mills, the source of Polartec® fabrics. After a severe fire, instead of closing down the business, he kept workers on salary during recovery—a widely acclaimed example of decency in business. http://www.knowltonmosaics.com/
- Peter Robinson kindly gave me permission to use the photo in Figure 6.3.23 of a sculpture of John Robinson. (John Robinson, Edition Limitee; http://www.johnrobinson.com/)

My thanks to The M. C. Escher Company-Holland for permitting me to use images of eight of Escher's artworks (Figures 2.6.4 through 2.6.6, 4.1.6, 4.5.7, 4.5.9, 5.1.25, and 6.3.5). All of these are copyrighted: © 2006 The M.C. Escher Company-Holland. All rights reserved. www.mcescher.com.

Thanks to Peter Sorensen for allowing me to use the photograph in Figure 2.3.15.

Figure 5.3.7 is from http://www.class.uidaho.edu/ngier/polyhedra/goc.htm. Permission to use it was granted by Nick Gier.

Thanks to the editor David Grubbs for his many prompt and very helpful technical suggestions and advice.

Chandler Davis, editor of *The Mathematical Intelligencer*, gave me permission to use the graphics shown is Figure 6.2.4.

This project was partially supported by The Winnipeg Foundation.

The following wise words by Madhusüdana Sarasvati, a sixteenth century Indian philosopher, are befitting this book:

Whatsoever the merits, they are not mine,
Whatsoever the faults, they are indeed, all mine.

SOME TECHNICAL ASPECTS

The character called Dadat was made with *PovRay*, as were some of the scenes. I have used *Electric Image Universe* to make some of the animations and some three-dimensional images. Almost all of the vector-images were done with *Mathematica* and then processed with *Adobe Illustrator*. My primary application for fractal images and fractal animations was *Escape 3.3*. Secondary adjustments in some of the images were done with *Adobe Photoshop*.

I used the program *Mathematica* extensively, especially in conjunction with various packages. Most of the hyperbolic tilings (Section 4.5) were produced with the package *Tess* by Miodrag Sremcevic and Radmila Sazdanovic. The books, *The Mathematica Guidebook for Graphics* by Michael Trott and *Exploring Mathematics with Mathematica* by Theodore W. Gray and Jerry Glynn supplied for me a few fine programs that I have used.

THE ACCOMPANYING CD

The enclosed CD contains a pdf file with the color figures used in this book, six short animations referred to in the book, as well as a folder called Extra containing more color figures and 15 additional short animations.

CHAPTER 1

Euclidean Geometry

The history of ages teaches us that virtually nothing in this world changes linearly, and that nothing expands uniformly in time. Human civilizations are no exception; there were times when they grew in their scope and depth, and there were dark epochs when calamities, catastrophes, and cataclysms left magnificent ancient cities in ruins. These ruins, sometimes buried under desert sand, sometimes covered by jungle vegetation, and sometimes inundated by rising seas and oceans, are silent witnesses of the existence of ancient civilizations with superior practical knowledge of rules of geometry (Figures 1.1.1 and 1.1.2).

According to orthodox archeology, the onset of modern civilization happened at about 3500 BC in Sumer, situated in the fertile lands between the rivers Tigris and Euphrates. There are relatively few written records left of the Sumerian civilization, but we do know that they possessed at least sophisticated practical knowledge of geometry. For example, the statement of (what we now call) the *Pythagorean theorem* was discovered on a Babylonian tablet dated long time ago, sometime around 1900 BC.

It was the ancient Greeks who made the next giant step in the direction of better understanding and interpreting the abstract and material universe. Greek philosophers were virtually obsessed with the beauty of geometrical objects and with their properties, and they found in geometry a perfect medium for abstract analysis of various phenomena.* The above-mentioned Pythagorean theorem is a case in point: Pythagoras was a Greek geometer (fifth century BC), mathematician, and philosopher, who realized that it is inadequate to accept the validity of properties just by empirical observations or measurements. For example, checking that the square over the hypotenuse of a right-angled triangle is the sum of the squares over the other two sides by simple measurements in many triangles is unsatisfactory, because it does not preclude the possibility of existence of some exotic triangle

---

* For example, according to Plato (as stated in his book *Timaeus*), the earth was made of tiny cubes, while the air was an aggregate of small tetrahedrons. We will discuss these objects in Chapter 5.

FIGURE 1.1.1 The Sphinx and the pyramid of Khafre, both of undetermined antiquity. Photo by Hajor (2002).

FIGURE 1.1.2 The temple of Quetzalcoatl, also known as the step pyramid of Chichen Itza.

where that property fails. Pythagoras wanted to understand *why* that property is satisfied. Thus, the notion of *proof* was born.

According to the Pythagorean theorem, $a^2 + b^2 = c^2$, where a, b, and c are the lengths of the sides of any right-angled triangle, with c being the length of the longest side. A visual proof is given in Animation 1.1.1 (CD) and illustrated Steps 1–6 below.

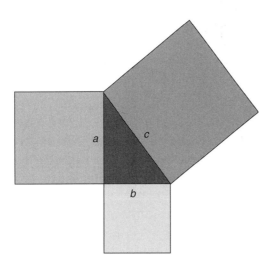

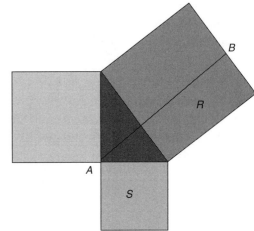

STEP 1. We want to show that the area of the largest square is the sum of the areas of the smaller squares.

STEP 2. Split the largest square into two rectangles as shown, then show that these rectangles have the same areas as the other two squares, respectively, and, in particular, that the area of R is the same as the area of S.

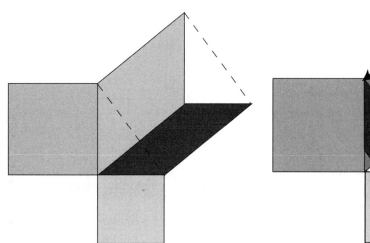

STEP 3. Shear the two rectangles along the line AB shown in Step 2 above. Shears do not affect areas.

STEP 4. Then rotate one of the parallelograms as indicated ....

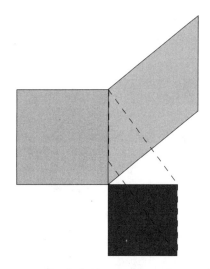

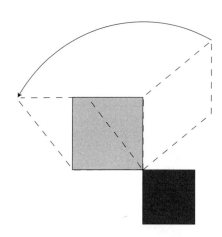

STEP 5. ... and shear again as shown. We have shown that one of the rectangles has the same area as the smallest square.

STEP 6. Finally, perform the same operations to show that the other rectangle has the same area as the remaining square.

We do not know whether the first proof belonged to Pythagoras, to his students, or perhaps to someone else, for their original work has been lost in the darkness of the bygone ages. What we do know is that the substance of the proof we give here exists in an ancient book, preserved by meticulous copying and recopying through thousands of years: Euclid's treatise *The Elements* (Book 1, Proposition 47).

We digress for a moment to prepare the background for the next exercise. Right-angled triangles with sides of integer lengths are called Pythagorean triangles. For example, a triangle with sides of lengths 3, 4, and 5 units is a (right-angled) Pythagorean triangle.

Exercise: Show that the height over the longest side of a triangle with sides of lengths 13, 14, and 15 units splits the triangle into two Pythagorean triangles.

1.1 THE FIVE AXIOMS OF EUCLIDEAN GEOMETRY

Euclid (Figure 1.1.3) lived and worked in Alexandria, a metropolis of the time, containing the largest ancient library (subsequently destroyed by various invaders). He was familiar

with the work of the so-called Athenian school of mathematics (see Figures 1.1.4 and 1.1.5), and a large part of the chapters on geometry in *The Elements* is a collected knowledge of that school. *The Elements*, especially the first two chapters containing Euclid's treatment of geometry, was the primary book of mathematics for many hundreds of years.

The importance of Euclid (as a geometer representing the school of Athens) is mainly in the fact that he went one step further than looking for proofs: he wanted to know what *proof* was—a fundamental question indeed. To explain what we mean by that, we now take a closer look at the visual proof of the Pythagorean theorem given above. It is

FIGURE 1.1.3 Euclid, about 325–265 BC.

FIGURE 1.1.4 Rafael. *School of Athens*, around 1500. The two men at the center of the picture depict Plato and Aristotle; Euclid is shown as the balding man among the group of students at the right bottom of the picture. We will deal with some of the Plato's work in Chapter 3. There is one more character visible in this picture: Dadat (a brief introductory note about him is given in Figure 1.1.5).

not hard to notice that beneath its surface there are claims of various degree of obviousness that we accept beforehand, and on which we rely. For example, in the above proof we accept that since the rectangle denoted by R in Step 2 has the same area as any one of the intermediate parallelograms, and since the latter have the same area as T, it follows that R and T have the same area. The underlying rule is one of the five **common sense axioms*** that Euclid explicitly listed in *The Elements*, and which he accepted without proof. Further, we implicitly assume in the proof that there is a line passing through the corner of the right angle in the triangle, and that is perpendicular to the hypotenuse of the

FIGURE 1.1.5 Dadat, about 20×10^9 BC to present. Everybody's contemporary.

starting triangle (see Step 2 above). Moreover, we also accept without formal justification that that line is perpendicular to the side of the rectangle R that is parallel to the hypotenuse of the starting triangle. Is that obvious? What is "obvious," and which claims deserve some kind of justification?

Euclid realized that one had to start with some reasonable rules of inference (mathematical thinking) and a set of acceptable claims NOT to be justified formally. That would set up a **theory**; only then could one proceed to deduce theorems. So, he set up a system of obvious claims that he called **axioms** or **postulates** (we will use both terms). Some of these axioms were common sense declarations (one example is given in the footnote below). Five of them were geometrical axioms, and we list these geometrical axioms as follows (in modernized wording and followed by illustrations and short explanations where needed):

1. For every point P and every point Q not equal to P, there is a unique line passing through P and Q (Figure 1.1.6).

FIGURE 1.1.6

2. For every segment AB and for every segment CD, there exists a unique point E such that B is between A and E, and the segment CD is congruent to the segment BE (Figure 1.1.7).

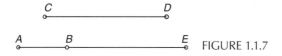

FIGURE 1.1.7

(This axiom tells us that every segment AB can be extended by any length whatsoever.)

3. For every point O and every point A not equal to O, there exists a circle with the center at O and with radius OA (Figure 1.1.8).

---

* Here is the first of Euclid's five *common sense axioms*: things that are equal to the same thing are also equal to one another.

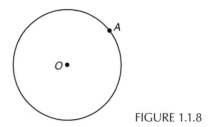

FIGURE 1.1.8

4. All right angles are congruent.

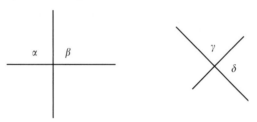

FIGURE 1.1.9

(A right angle is defined as being any of the four angles determined by two intersecting lines with equal supplementary angles. If the supplementary angles α and β in Figure 1.1.9 are equal, then both are right angles. This axiom tells us that if $\alpha = \beta$ and if $\gamma = \delta$, so that all these angles are right angles, then we can move, say, the angle α, so that it exactly overlaps γ.)

5. For every line l and every point P that does not lie on l, there exists a unique line m through P and parallel to l (Figure 1.1.10).

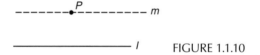

FIGURE 1.1.10

In *The Elements*, Euclid extracted 28 theorems from his axioms, and his treatise was to become the foremost book of geometry and mathematics during the next two millennia or more. Various mathematicians of ancient, medieval, and modern times expanded Euclid's work by adding many more theorems, whereas some others tried to improve on Euclid's theory. The majority of the latter group tried hard to simplify Euclid's theory by showing that some of the axioms could in fact be proven from the other remaining axioms (in which case the former would have been redundant in the theory, and so could have been eliminated as axioms and listed as theorems). Most of their attention was focused on the last, fifth axiom: many people felt that this axiom could somehow be proven or deduced from the previous four axioms, and some of these thinkers spent years trying to do that. This endeavor came to a very unexpected conclusion sometime around the middle of nineteenth century. We will not reveal the outcome until Chapter 4.

1.2   RULER AND COMPASS CONSTRUCTIONS

Constructing geometrical objects was of great interest to ancient civilizations, particularly to the ancient Greeks. They were interested in this subject not only from the philosophical point of view, but also because of practical applications, mostly in architecture.

The first and third axioms of the Euclidean geometry stipulate constructibility of a line through two distinct points, and a circle centered at a point and with a given radius, respectively. What else can be constructed based on these two axioms, as well as on the other three axioms? This was a central problem during the two millennia following Euclid, and many people devoted their lifetimes to discovering and exploring various constructions. Our goal will be to present some of the most basic of these constructions.

Since an unmarked ruler (a straightedge) is needed to construct a line segment through two given points (Axiom 1), and since we need a compass to draw a circle at a given center and with a given radius (Axiom 3), it was natural to focus on constructions involving only these two tools (now called Euclidean tools). In this section, we explore various constructions that can be done using only an unmarked ruler and a compass. By the end of it, we will indicate some of the limitations of the Euclidean constructions. We point out, however, that we will not define precisely what we mean by "using an unmarked ruler and a compass"; instead we will rely on the examples and on our intuition.

Before we begin we make a short note regarding our notation: if A and B are the end points of a line segment, then we will use AB to denote that line segment, as well as the length of that line segment. It should be clear from the context which of the two meanings is being used.

CONSTRUCTION 1. BISECTING A LINE SEGMENT AND CONSTRUCTING A PERPENDICULAR

We are given a line segment; we want to construct a line passing through the middle of that line segment and perpendicular to the line segment.

Solution: This is one of the most basic constructions and we describe it in illustrated Steps 1–3 as follows:

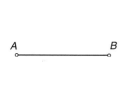 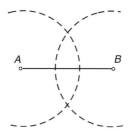 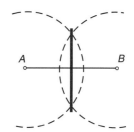

STEP 1. Given a line segment AB …

STEP 2. … draw any two intersecting circles of equal radii and centered at the points A and B.

STEP 3. Find the points of intersection of the circles and draw the line through them. □

CONSTRUCTION 2. CONSTRUCTING A LINE THROUGH A GIVEN POINT AND PERPENDICULAR TO A GIVEN LINE

Suppose l is a line and C is a point out of l. Construct a line through C and perpendicular to l.

Solution: As explained in the caption of Step 1, we reduce this construction to the one done in Construction 1.

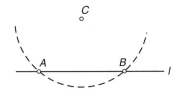

STEP 1. Draw a circle centered at the given point C and large enough to intersect the line l at two points, denoted by A and B. Then proceed exactly as in Construction 1, constructing the line bisecting the line segment AB. That line must pass through C (since bisectors of chords of a circle pass through its center), and it is by construction perpendicular to l. □

CONSTRUCTION 3. DUPLICATING AN ANGLE

Given an angle (denoted α in the picture), construct an angle with one of the sides over the given line l and of the same size as α.

Solution:

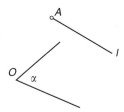

STEP 1. Given an angle, a line l, and a point A on the line.

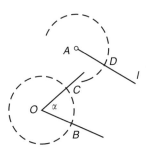

STEP 2. Draw a circle centered at O and of any radius. Draw the circle centered at A with the same radius. Identify the intersection points B, C, and D.

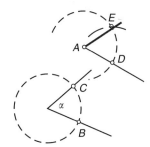

STEP 3. Construct the circle centered at D and with radius equal to BC. Find the intersection point E and join A to E. □

CONSTRUCTION 4. DRAWING A PARALLEL LINE

Construct the line passing through a point A and parallel to a given line l.

Solution:

$_\circ^A$

——————— l STEP 1. Given a line l and a point A out of the line.

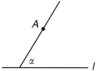

STEP 2. Draw any line through A intersecting l and identify the angle α.

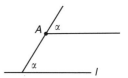

STEP 3. Duplicate the angle α at A as shown, so that the line constructed in the previous step is one of its two sides. [*Note:* the half lines making the sides of an angle are sometimes called angle brackets.]

Note that the line through A in the second step above can be chosen arbitrarily (as long as it intersects l). The above construction is somewhat simpler if that line is such that the angle α is 90°. But, then again, if we want to start with such a line, we would have to construct it (see Exercise 2). □

CONSTRUCTION 5. IDENTIFYING THE CENTER OF A PARTIALLY GIVEN CIRCLE

We are given three points on a circle. Find the center of the circle and draw the circle.

Solution:

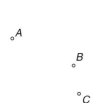

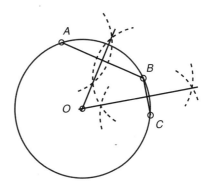

STEP 1. Given three points A, B, and C, not on the same line.

STEP 2. Bisect the line segments AB and BC (see Construction 1). The intersection point O of the bisecting lines is the center of the circle. The radius is any of OA = OB = OC. □

CONSTRUCTION 6. CONSTRUCTING A CIRCLE CENTERED AT A GIVEN POINT AND TOUCHING ANOTHER CIRCLE

We will first give an *incorrect* solution of the problems stated in the subtitle. It contains a common error that will help us further clarify the use of compass as an Euclidean tool.

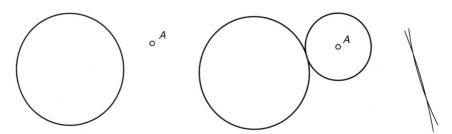

AN INCORRECT CONSTRUCTION: The setup is given to the left (a circle and a point outside the circle). We center the compass at the given point and open it so that the circle of that radius touches the given one (middle illustration). The problem of this approach is illustrated to the right: after zooming 10-fold we see that the two circles do not touch after all.

As noted in the caption of the last illustration, we centered the compass at A and opened it just as much so that the circle we draw touches the given circle. The problem is to determine precisely how much is "just as much." As long as that problem is unsolved we would never be sure that zooming in by many times around the point where we expect the circles to touch would reveal that these circles do in fact intersect or have no common points at all.

We might, then, do an adjustment (and decrease the radius of the new circle by a bit), but the problem will persist (perhaps zooming again would reveal another flaw).

Solution: Fortunately, there is an easy way out of this predicament: we can first precisely determine the radius of the circle we want to draw, and then we can proceed to construct the circle. The details are given in illustrated Steps 1–3 below.

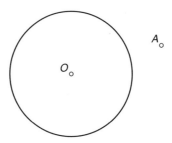

STEP 1. First identify the center O of the given circle. See Construction 5 above.

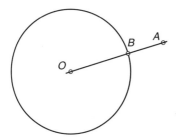

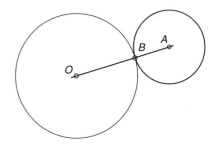

STEP 2. Then join the points A and O and find the intersection point B: this is where the two circles should touch.

STEP 3. Finally, draw the circle centered at A and with radius exactly equal to AB.
□

Exercise: Observe that there is one more, larger circle that touches the given circle and centered at the given point. Construct it.

CONSTRUCTION 7. SUBDIVIDING A SEGMENT INTO EQUAL PARTS

In the example below, we subdivide the given segment into four equal parts. A very similar construction is needed for any number of equal parts.

Solution:

STEP 1. We are given a line segment AB.

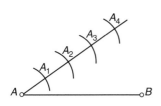

STEP 2. Draw any line through A not overlapping with AB. Then mark points A_1, A_2, A_3, and A_4 using a compass opened at the same fixed radius (so that all the small segments are equal in length).

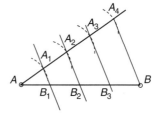

STEP 3. Connect A_4 with B, then construct lines parallel to A_4B and through A_3, A_2, and A_1, respectively (as in Construction 4). The intersection points of these three lines with AB give the sought points of subdivision. ☐

A Bit of Math. Justifying Construction 7

Let us see why, say, $AB_1 = B_1B_2$. The triangles AA_1B_1 and AA_2B_2 are similar, and so their sides are proportional. Since, by construction, AA_2 is twice longer than AA_1, it follows that AB_2 is twice longer than AB_1. So, $AB_1 = B_1B_2$. ☐

CONSTRUCTION 8. MULTIPLICATION OF NUMBERS

We show how to multiply numbers geometrically. More precisely, given two line segments, one of length m and one of length n, we will see how to construct a line segment of length mn (m times n), assuming that a line segment of unit length is given.

Solution:

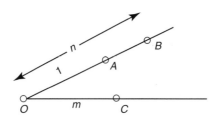

STEP 1. Plot the given data as in the picture: $OA = 1$, $OB = n$, and $OC = m$.

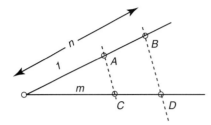

STEP 2. Connect A and C and then construct the line through B that is parallel to AC (as in Construction 4). The intersection of that line and the line through O and C is denoted by D. The length of OD is exactly n times m units.

We have implicitly assumed above that $n > 1$. This is not essential and the same idea works if $n \leq 1$. ☐

A Bit of Math. Justifying Construction 8

Here is why $OD = mn$. The triangles OAC and OBD are similar since they have the same corresponding angles. Thereby their sides are proportional. Denoting $OD = x$, the similarity of the triangles implies that $\frac{n}{1} = \frac{x}{m}$, from where we find that $x = mn$ as claimed. ☐

Exercise: Modify the above construction to the case when both n and m are less than 1. (What is to be done if either n or m is equal to 1?)

The Story of Three Ancient Problems

Here is an innocent-looking construction. We start with a right angle *POR* (see the illustration below), and with a point *A* in the angle *POR*. First we construct the rectangle with *O* and *A* being the opposite vertices as shown in the illustration, and then we identify the point *E* in the intersection of the diagonals of that rectangle. (All that can be done using either Construction 2 or Construction 4.) Our goal is to construct the line through *A* such that the intersection points *B* and *C* of that line with the rays *OR* and *OP*, respectively, are equidistant from the point *E* (that is, such that *EB* = *EC*). One attempt to solve this problem is outlined in the caption of the illustration.

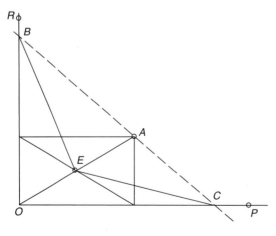

HERON'S CONSTRUCTION Position the edge of a ruler through A and then turn it about A until EB = EC.

There is a major problem here—turning about *A as much as needed* to get *EB = EC* is *not* a Euclidean construction, and this step does not conform to our (implicit) definition of *using a ruler*. We did manage to avoid the *wiggling* compass in Construction 6. However, in this case of a *wiggling ruler*, there is no easy way out: in this case it is *not* possible to construct, by means of an unmarked ruler and a compass, a line through *A* such that *EB = EC*.

We note in passing that it is possible to construct, using a ruler and a compass, a line through *A* such that *OB = OC* (Exercise 9), and a line through *A* such that *AB = AC* (Exercise 10).

The general problem of showing that some constructions are not possible proved to be much more difficult than the problem of finding constructions when they do exist. One of the main obstacles to overcome by people in the pursuit of solutions of such problems was purely psychological: the idea that a construction could not be done no matter how hard one tried was not an easy prospect to accept.

The following three problems are called the three geometric problems of antiquity (in all of them the word *construct* means, of course, construct by using an unmarked ruler and a compass).

Ancient Problem 1 (squaring a circle): Construct a square with area same as a given circle.

Ancient Problem 2 (doubling a cube): Construct a cube of volume twice the volume of a given cube.

Ancient Problem 3 (trisecting an angle): Trisect an angle (subdivide it into three equal angles).

Heron, a Greek geometer who lived in the first century, tried to solve the Ancient Problem 2 by constructing the line through *A* such that *EB = EC*. It could be shown that such a construction (but without using a *wiggling* ruler) implies a solution to Ancient Problem 2.

Beside Heron, literally thousands of people, from ancient Greeks, through Arabs in the medieval world, to western thinkers and mathematicians, tried their best to solve these problems. No one could. The endeavor of searching for *right* constructions was brought to an abrupt and unexpected end in the nineteenth century, after a few seemingly nonrelated and purely algebraic results led to the proof that none of these three problems is solvable.

Exercises: In the following problems the word "construct" is short for "construct using an unmarked ruler and a compass."

1. Construct an equilateral triangle over a given side (this yields another construction of an angle of 60°).
2. Given a point *A* on a line *l*, construct a line through *A* and perpendicular to *l*.
3. Bisect a given angle.
4. Double a given angle.
5. Construct an angle of 15°.
6. Subdivide a line segment into seven equal parts.
7. Given a line *l* and a point *A* outside *l*, construct a line *m* passing through *A* and intersecting *l* at 30°.

8. a. Given one of the two diagonals of a square construct the square.
   b. Given a segment of length d construct a segment of lengt $\sqrt{d}$.
9. Given an angle (two lines intersecting at a point O) and an arbitrary point A, construct a line passing through A and intersecting the two given lines at the points B and C in such a way that $OB = OC$.
10*. Given an angle (two lines intersecting at a point O) and an arbitrary point A, construct a line passing through A and intersecting the two given lines at the points B and C in such a way that $AB = AC$.
11*. Given two line segments, one of length n, and the other of length m (n and m are arbitrary positive numbers), construct a line segment of length m/n. Assume you have a line segment of unit length.

1.3 THE GOLDEN RATIO

Ancient Greek mathematicians were philosophers and artists at the same time, and elegance and other esthetic aspects of their work mattered to them. According to these ancient sages, a major part of the elegance of objects was *proportionality*; they defined what that meant to them very explicitly, at least in case of geometric objects.

$$A \qquad\qquad\qquad C \qquad\qquad B$$

If a point C on a line segment AB is such that $\dfrac{AB}{AC} = \dfrac{AC}{BC}$, then ratio AB/BC (or AC/BC) is called the **golden ratio**, or the **golden proportion**, or the **golden mean** (we will stick with the first term). The point C is called a **golden cut** of the line segment AB. As we see in the math note below, the golden ratio is always equal to $\dfrac{1 + \sqrt{5}}{2}$ (approximately equal to 1.61803) and it does not depend on the length of the original line segment AB. The number $\dfrac{1 + \sqrt{5}}{2}$ is traditionally denoted by the Greek letter ϕ (phi).

A Bit of Math. The Value of the Golden Ratio
We check below that the golden ratio $\dfrac{AB}{AC}$ is indeed $\dfrac{1 + \sqrt{5}}{2}$. Denote it (for easier manipulation) by x, and denote $AB = a$, $AC = y$ (so that $BC = a - y$). So, $\dfrac{a}{y} = x$, and for the time being we know that it is a positive number. The equation $\dfrac{AB}{AC} = \dfrac{AC}{BC}$ becomes $\dfrac{a}{y} = \dfrac{y}{a-y}$, or $\dfrac{a}{y} = \dfrac{1}{\frac{a-y}{y}}$, or $\dfrac{a}{y} = \dfrac{1}{\frac{a}{y} - 1}$. Since $\dfrac{a}{y} = x$, the last equation is equivalent to (means the same as) $x = \dfrac{1}{x - 1}$, or $x(x - 1) = 1$, or (after multiplying and rearranging) $x^2 - x - 1 = 0$. The quadratic equation then gives us two solutions: $x = \dfrac{1 - \sqrt{5}}{2}$ and $x = \dfrac{1 + \sqrt{5}}{2}$. The first one is negative, and so we disregard it. The second solution is postive and so we can conclude that the only solution that fits our data is $x = \dfrac{1 + \sqrt{5}}{2}$. □

CONSTRUCTION 1. A GOLDEN CUT

In illustrated Steps 1–3 we show how to construct a golden cut C of a line segment AB.

[*Note*: The word "construct" here and in the other examples and exercises in this section means, as it is a standard by now, "construct using an unmarked ruler and a compass."]

Solution:

STEP 1. We are given a line segment.

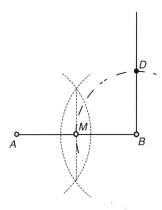

STEP 2. Find the midpoint M of AB, then construct (as in Construction 2, Section 1.2; we do not show that part of construction here) the perpendicular to AB at B. Then mark the point D such that $BM = BD$.

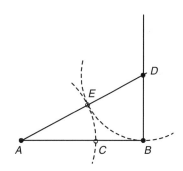

STEP 3. Draw a circle with radius BD and centered at D and find the intersection point E of that circle with the segment AD. Draw another circle centered at A and with radius AE; the intersection of this circle with AB is the desired golden cut C. ☐

A Bit of Math. Justifying the Construction of the Golden Cut

Denote $AB = a$. Since DE is (by construction) of length $\frac{a}{2}$, it follows that $AC = AE = AD - \frac{a}{2}$. The length AB can be found by the Pythagorean theorem (see the triangle ABD): $AD = \sqrt{a^2 + \left(\frac{a}{2}\right)^2}$. Using this and simplifying we obtain $AE = a\frac{\sqrt{5}}{2} - \frac{a}{2}$, so that $AC = a\frac{\sqrt{5}}{2} - \frac{a}{2}$ too. Finally we compute $\dfrac{AB}{AC} = \dfrac{a}{a\frac{\sqrt{5}}{2} - \frac{a}{2}} = \dfrac{1}{\frac{\sqrt{5}}{2} - \frac{1}{2}} = \dfrac{2}{\sqrt{5} - 1} = \dfrac{2}{\sqrt{5} - 1}\dfrac{\sqrt{5} + 1}{\sqrt{5} + 1} = \dfrac{2(\sqrt{5} + 1)}{4} = \dfrac{1 + \sqrt{5}}{2}$, as claimed. ☐

In Figure 1.3.1, we see the ancient building of the Parthenon in Athens. The rectangle held by Dadat and those superimposed on the image of the Parthenon are **golden rectangles**, which means that the ratio of the base of the rectangle over its height is exactly the

FIGURE 1.3.1 Parthenon, Athens. Built 448–432 BC. Golden rectangles abound.

golden ratio. Notice how the Parthenon was built to conform to the ancient Greek concepts of proportionality.*

FIGURE 1.3.2 The Great Pyramid of Khufu in Giza. Photo by Nina Aldin Thune (2005).

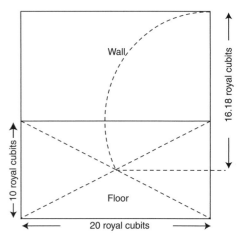

FIGURE 1.3.3 The king's chamber: the golden ratio appears!

It is not at all clear whether the ancient Greeks were the first to identify and use the notion of the golden ratio. In order to support (but not to prove) this claim, we will now retrieve the golden ratio from within a much older archeological structure. Consider the Great Pyramid of Khufu in Giza (Figure 1.3.2), one of the seven wonders. A blueprint of the king's chamber, a compartment in the interior of the Great Pyramid, is sketched in Figure 1.3.3. Notice the number 16.18!

Is the emergence of the golden ratio within the dimensions of the king's chamber a coincidence? The answer to this question depends on the interpretation of the choices made

---

* Not all historians agree that the golden ratio was consciously used in the construction of the Parthenon. Indeed, it is a valid skeptical proposition that one can superimpose the golden rectangle upon any *sufficiently complicated* structure in such a way that it fits some rectangular portion of that structure.

for the dimensions of the chamber, because once the height of the chamber was chosen to be one-half of the diagonal of the rectangular base, then the golden ratio was bound to appear. This claim is clarified in Exercise 9a.

CONSTRUCTION 2. A GOLDEN RECTANGLE OVER A GIVEN HEIGHT

Solution:

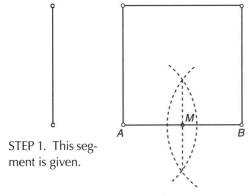

STEP 1. This segment is given.

STEP 2. Construct a square over the given segment, then find the middle point M of the base AB.

1 : 1.6.

STEP 3. Open the compass from M to C and draw the circle to get the intersecting point D. Then segment AD is the base of the rectangle we wanted. □

Question. Take a look at the smaller rectangle *BDEC* in the above construction. Why is it a golden rectangle too?

A Bit of Math. Justifying the Construction of the Golden Rectangle

Let us show that the rectangle we have just constructed is indeed golden. We need to show that the ratio of the base over the height of the rectangle is $\frac{1 + \sqrt{5}}{2}$. Denote the length of the given segment by a. So, $AB = BC = a$ and $MB = a/2$. By applying Pythagorean theorem to triangle MBC, we obtain $MC = \sqrt{\left(\frac{a}{2}\right)^2 + a^2}$, which, after simplification, becomes $a\frac{\sqrt{5}}{2}$. By construction (the illustration in Step 3), MD is also equal to $a\frac{\sqrt{5}}{2}$. So, the base AD is $AM + MD$, which, as we have found out, is equal to $\frac{a}{2} + a\frac{\sqrt{5}}{2}$. Consequently, the ratio AD over BC is $\frac{\frac{a}{2} + a\frac{\sqrt{5}}{2}}{a}$. After canceling, this becomes exactly $\frac{1 + \sqrt{5}}{2}$ as claimed. □

Digression. According to a study,* people's perception of beauty of a human face is closely related to whether or not the face fits nicely into a (vertical) golden rectangle.

CONSTRUCTION 3. A GOLDEN SPIRAL

Solution: As we have noticed in Construction 2, if we construct the square over the smaller side in a golden rectangle and within the golden rectangle, then the leftover rectangle is also a golden rectangle. We use this observation in the next construction.

STEP 1. Start with a golden rectangle.

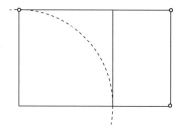

STEP 2. Construct square as shown.

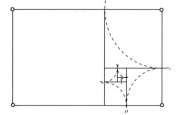

STEP 3. Do that again inside the smaller leftover golden rectangle. Continue the process.

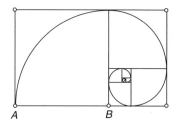

STEP 4. Once we have a sequence of smaller and smaller squares, inscribe quarter circles in each of them as shown: for example, the first quarter circle is centered at *B* and has radius *AB*.

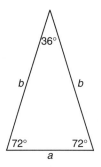

FIGURE 1.3.4 An acute golden triangle: *b/a* is the golden ratio.

Golden Triangles

An isosceles acute triangle is a ***golden triangle*** if the ratio of the length of the side over the length of the base is the golden ratio (Figure 1.3.4).

It can be shown that the interior angles of any acute golden triangle are 72°, 72°, and 36°, as indicated in Figure 1.3.4 (see also Exercise 11(a)). This relationship extends in the opposite direction too: any isosceles triangle with these interior angles must be an acute golden triangle. This observation simplifies the following constructions.

* In 1994, orthodontist Mark Lowey made detailed measurements of fashion models' faces. He asserted that the reason we classify certain people as beautiful is because they come closer to golden ratio proportions in the face than the rest of the population. (Found in *The Golden Ratio and Aesthetics* by Mario Livio, plus.maths.org/issue22/.)

CONSTRUCTION 4. AN ACUTE GOLDEN TRIANGLE OVER A GIVEN BASE

Solution: The idea is simple: Construct an acute golden triangle of arbitrary size and then use its angles to get the one wanted.

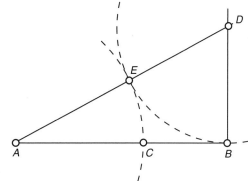

STEP 1. Start with the given base (and we forget it until the last step).

STEP 2. Now take any line segment *AB* and repeat Construction 1 in this section to get the golden cut *C*.

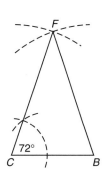

STEP 3. Construct an isosceles triangle over *CB* and with the other two sides equal to *AC*. The angle at *C* must be 72°.

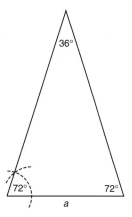

STEP 4. Duplicate the angle of 72° at one of the end vertex of the given base *a*, and do the same at the other end.   □

CONSTRUCTION 5. AN ALTERNATE CONSTRUCTION OF AN ACUTE GOLDEN TRIANGLE OVER A GIVEN BASE

Solution: Construct a golden rectangle with the line segment *a* being its smaller side (as in Construction 2 above). The longer side of the rectangle is the same as each of the two equal sides of the golden triangle we are constructing. Thus, we have the lengths of all sides of the golden triangle. The rest is simple enough and we leave it to the reader.   □

Closely related to acute golden triangles are obtuse golden triangles: they are also isosceles triangles, but this time the ratio of the length of the base with respect to the length of any of the other two sides is the golden ratio.

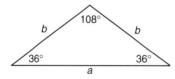

FIGURE 1.3.5 An obtuse golden triangle: this time *a*/*b* is the golden ratio.

The angles in all obtuse golden triangles are also fixed: they must be 36°, 36°, and 108°, as indicated in Figure 1.3.5 (see Exercise 11(b)).

CONSTRUCTION 6. SUBDIVIDING OBTUSE GOLDEN TRIANGLES

As illustrated in Step 1, we see how an obtuse golden triangle can be subdivided into one acute and one obtuse triangle.

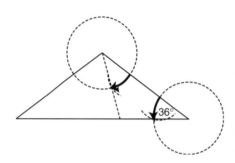

STEP 1. We simply duplicate the angle of 36° at the top vertex (see the details of construction in Section 1.2). It is easy to show that the interior angles of the two smaller triangles are as they should be for them to be golden triangles; we leave it as an exercise (Exercise 2). □

An application of Construction 6 is shown in Figure 1.3.6. The spiral is composed of arcs over the largest sides of consecutive obtuse golden triangles. More details are given in Exercise 3.

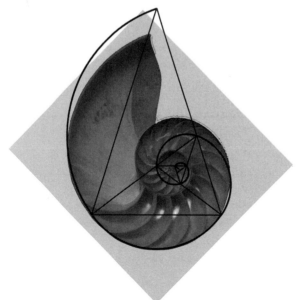

FIGURE 1.3.6 A shell superimposed over a golden spiral over an acute golden triangle.

Regular Pentagons

A ***regular pentagon*** is a convex (ignore this stipulation for a while) polygon with five equal sides and five equal interior angles (Figure 1.3.7).

Since all sides are equal, it follows from symmetry that the central angles, denoted by α in Figure 1.3.8, are also equal. Since the full circle makes an angle of 360°, it follows that $\alpha = \dfrac{360°}{5} = 72°$.

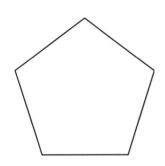

FIGURE 1.3.7 A regular pentagon.

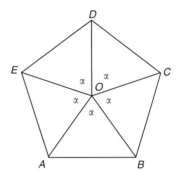

FIGURE 1.3.8

We recall that two of the angles in every acute golden triangle are 72°. Is the triangle ABO a golden triangle? The answer is negative: the other two angles are of size (180°–72°)/2, which is equal to 54°. However, there are golden triangles lying hidden within the pentagon, and we uncover them in Figure 1.3.9. Notice that the interior angle at the vertex A (and so, at any other vertex) of the pentagon is twice 54°, that is, 108°. So, the angle at the vertex C of the pentagon is 108°. It follows that the angle at B in the triangle BCD is equal to (180° − 108°)/2, that is, 36°. Consequently, this triangle is an obtuse golden triangle. By symmetry, the triangle ADE is also an obtuse golden triangle. As we expect by now, the acute triangle in the middle is an acute golden triangle. This is true since the interior angles at A and B are both 108° − 36° = 72° (108° is the interior angle at B of the pentagon, 36° is the angle at B of the triangle BCD—both these facts were observed above), and it follows that the angle at D of this acute triangle is 36°.

Summarizing, every regular pentagon can be subdivided into two obtuse golden triangles and one acute golden triangle. This is what we need to justify in the following construction.

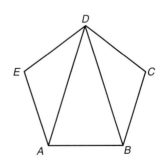

FIGURE 1.3.9

CONSTRUCTION 7. A REGULAR PENTAGON OVER A GIVEN SIDE

Solution:

STEP 1. First construct a golden acute triangle over the given base *AB* (as in Construction 4).

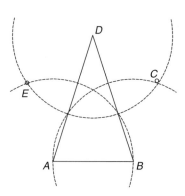

STEP 2. Then draw three circles centered at *A*, *B*, and *D*, respectively, and with radius *AB*. Find the intersecting points *C* and *E*.

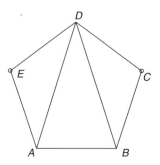

STEP 3. Connect *E* with *A* and *D*, and connect *C* with *B* and *D*. *ABCDE* is a regular pentagon.

□

A Bit of Math. A Formula for the Golden Ratio

We show that $\phi = \sqrt{1+ \sqrt{1+\sqrt{1+ \cdots}}}$, where it is assumed that the pattern in the expression within the square root never ends. Precisely because this pattern is never ending, we notice that the expression within the outermost root is 1 plus ϕ. Denoting $x = \sqrt{1 + \sqrt{1 + \sqrt{1 + \cdots}}}$, we have $x = \sqrt{1 + x}$. Starting with $x = \sqrt{1 + x}$, we square both sides to obtain $x^2 = 1 + x$, or $x^2 - x - 1 = 0$. We have already solved this quadratic equation, and of the two solutions we obtained the positive one is $\dfrac{1 + \sqrt{5}}{2}$, that is, the golden ratio*.

□

---

* There is a fairly large gap in our proof that $\phi = \sqrt{1+\sqrt{1+\sqrt{1+ \cdots}}}$! We have implicitly assumed that the expression $\sqrt{1+\sqrt{1+\sqrt{1+ \cdots}}}$ makes sense (i.e., it indeed defines a number). This claim is true, but justifying it goes beyond our scope. Such problems of existence are usually covered in intermediate calculus courses.

Exercises: As before, "construct" is short for "construct using an unmarked ruler and a compass."

1. Construct an obtuse golden triangle over a given base.
2. Show that in Step 1 of Construction 6, the dotted line segment indeed subdivides the large obtuse golden triangle into two smaller golden triangles. [*Hint*: show that the interior angles of the small obtuse triangle are 36°, 36°, and 108°, whereas the interior angles of the small acute triangle are 72°, 72°, and 36°.]
3. The first two steps in the construction of a golden spiral starting from an obtuse golden triangle are shown in Figure 1.3.10 (the two arcs shown in full line). Do the next 3 steps in the construction (by repeated subdivision of the obtuse triangles into pairs of acute and obtuse triangles and drawing smaller and smaller arcs around the obtuse triangles).

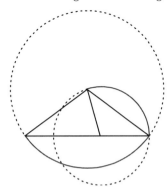

FIGURE 1.3.10 Keep in mind that only the obtuse triangles are subdivided in this process, and that we get the arcs making the spiral by repeatedly drawing circles centered at the vertex corresponding to the interior angles of 108° and passing through the ends of the bases of the small obtuse triangles. The acute golden triangles are just a side effect in this construction.

4. Start with an acute golden triangle and subdivide it into a pair of once acute and one obtuse smaller triangles.
5. Construct a golden rectangle over a given base (the longer side is given). [*Hint*: find the golden cut and then take a close look at the longer of the two segments.]
6. Where should we choose the point C on the line segment AB so that the rectangle $ABDE$ has the same area as the square $ACFG$ (see Figure 1.3.11)? Justify your answer.

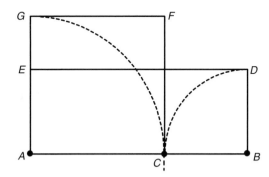

FIGURE 1.3.11

7. You are commissioned to construct an acute golden triangle using copper wire. If the total length of the wire is 4236 ft., what is the base of the largest acute golden triangle that you can construct? Draw a diagram to explain your answer.
8. In the star diagram shown in Figure 1.3.12 the shaded region is a regular pentagon. Find the value of the angle measure β. Give reasons for your answer.
9. a. With h and a as indicated in Figure 1.3.13, show that $(h + a/2)/a$ is the golden ratio. (Also see Figures 1.3.2 and 1.3.3.)

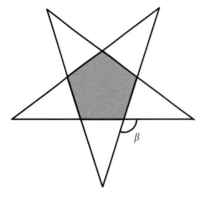

FIGURE 1.3.12

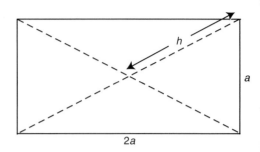

FIGURE 1.3.13

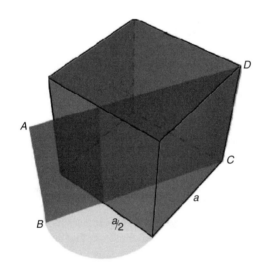

FIGURE 1.3.14

    b. Show that the rectangle *ABCD*, obtained from a cube of side *a* and a circle of radius *a/2* as shown in Figure 1.3.14, is a golden rectangle.

10. Show that the infinite expresson $1 + \cfrac{1}{1 + \cfrac{1}{1 + \cfrac{1}{1 + \cdots}}}$ is equal to the golden ratio. (You may assume that the expression defines a number.)

11. a. Show that every acute golden triangle $\triangle ABC$ with a base *AB* has angles of 72°, 72° and 36° at the vertices *A, B,* and *C* respectively. (*Hint*: see Exercise 4.)

    b. Show that every obtuse triangle $\triangle ABC$ with a base *AB* has angles of 36°, 36° and 108° at the vertices *A, B,* and *C* respectively.

1.4 FIBONACCI NUMBERS

From the times of ancient Greeks we fast-forward to the middle ages, sometime around the year AD 1200. At that time, 30-year old Leonardo Pisano Fibonacci (Figure 1.4.1), who sometimes called himself Bigollo (good-for-nothing, or traveler), returned to Pisa from North Africa and started to hand-write his first book, *Liber Abaci*. The book was published

2 years later. The third section of this book contains the following problem:

> *A certain man put a pair of rabbits in a place surrounded on all sides by a wall. How many pairs of rabbits can be produced from that pair in a year if it is supposed that every month each pair begets a new pair, which from the second month on becomes productive?*

Fibonnaci assumed implicitly that no rabbit dies.

Leonardo Pisano, of the wealthy family of Bonacci in Pisa, is now known under the name Fibonacci (short for "the son of Bonacci"). The increasing sequence of numbers of pairs of rabbits is now called the **Fibonacci sequence**, and the numbers in this sequence are called the **Fibonacci numbers**.

FIGURE 1.4.1 Leonardo Pisano Fibonacci, AD 1170–1250.

Let us briefly examine this sequence. For future reference, denote the number of pairs of rabbits after n months by f_n. Initially there is one pair of baby rabbits, so that $f_1 = 1$. They are not yet productive after one more month, so that f_2 is also 1. The third month the original pair produces its first pair of baby rabbits, so $f_3 = 2$. The fourth month the original pair produces one more pair of baby rabbits (while the other pair are too young for that yet); we count $f_4 = 3$. Next month we have two mature pairs which reproduce, so that the total after 5 months is 3 (old pairs of rabbits) + 2 (pairs of baby rabbits), a total of $f_5 = 5$ pairs. We make one more step: after 6 months we need to add $f_5 = 5$ (the total the previous month) to $f_4 = 3$ (the number of breeding, older pairs of rabbits from 2 months earlier, each of them producing one new pair of baby rabbits) to get $f_6 = 8$.

From the above analysis of the numbers during the first few months a formula emerges: the number (f_n) of pairs of rabbits after n months is equal to the sum of the number (f_{n-1}) of pairs of the previous month and the number of new pairs of baby rabbits (which is equal to the number of breeding pairs), which in turn is equal to the number of at least 2-month old pairs of rabbits, that is, f_{n-2}. In other words, $f_n = f_{n-1} + f_{n-2}$.

We summarize what we have found above: $f_1 = 1, f_2 = 1, f_3 = 2, f_4 = 3, f_5 = 5, ..., f_n = f_{n-1} + f_{n-2}$. This is a special way of describing numerical objects: we say that the Fibonacci numbers have been defined **recursively**. Our formula falls short of describing the number of rabbits f_n explicitly, but, nevertheless, it does provide in principle a method of determining any f_n. For example, in order to find f_{100}, we first need to establish the values of f_{99} and f_{98}; the latter needs f_{97} and f_{96}, then f_{96} is determined by f_{95} and f_{94}, and so on; eventually, we must back all the way to the small numbers we have established above, and then trace our way in the opposite direction until we reach f_{100}.

Is there an explicit formula for f_n? Yes there is! The rather surprising formula will be revealed (but not derived) later in this section.

Exercise: How many rabbits are produced after 9 months? (That is, find f_9.)

FIGURE 1.4.2 Thirty-Four Fibonacci rabbits and their keeper. Exercise: after how many months since the initial pair of rabbits had been introduced was this "photograph" taken?

You may have asked yourself by now "how is this related to the material we have covered so far in this chapter?" Before we give an answer to this question, let us take a look at some flowers.

Flowers, Fibonacci Numbers, and the Golden Ratio

One common property of these three beautiful plants (see Figure 1.4.3) is that the seeds are distributed in spiraling patterns: a set of spirals winding clockwise, the other set winding

(a)   (b)   (c)

FIGURE 1.4.3 Flowers.

FIGURE 1.4.4 Spirals winding clockwise.

FIGURE 1.4.5 Spirals winding counter clockwise.

counterclockwise. Let us pay closer attention to these two sets of spirals; the cactus seems the simplest and has the most visible spirals, so we focus on it.

There are exactly 13 spirals winding in the clockwise direction (Figure 1.4.4) and 21 spirals in the counterclockwise direction (Figure 1.4.5). Are these numbers of any significance? Consider again the first few Fibonacci numbers: $f_1 = 1$, $f_2 = 1, f_3 = 2, f_4 = 3$, and $f_5 = 5$. Continue a few more steps: $f_6 = f_4 + f_5 = 3 + 5 = 8$, $f_7 = f_5 + f_6 = 5 + 8 = 13, f_8 = f_6 + f_7 = 8 + 13 = 21$. And, there they are: the numbers of clockwise spirals and counterclockwise spirals are two consecutive numbers in the Fibonacci sequence of numbers. Coincidence? Hardly! The numbers of clockwise and counterclockwise spirals of the other two flowers are also two consecutive Fibonacci numbers. What is going on here?

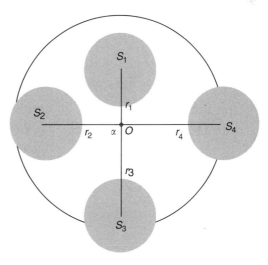

FIGURE 1.4.6 The first four seeds of young cyber-flower.

To come closer to the answer we resort to a cyber-flower. Here is a cyber-sunflower magnified at the center (Figure 1.4.6).

The first seed (S_1) to propagate in a sunflower is the closest to the center O. Then comes the next, S_2, some larger distance (r_2), away from the center. The third seed (S_3), is still farther from the center. The radial **angle of displacement** between two consecutive seeds is always a fixed angle α (Figure 1.4.6), while the distance from the center increases in some way, depending on the flower. In the first few examples (Figures 1.4.6 through 1.4.10), we consider that the distance between the nth seed and the center of the flower (the radial

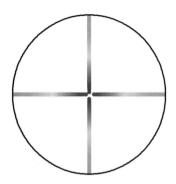

FIGURE 1.4.7 The Flower in Figure 1.4.6 will mature into a mutant flower.

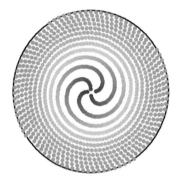

FIGURE 1.4.8 A cyber-flower with a displacement radial angle of 91°.

distance) is $(0.01)\sqrt{n}$ units. Different flowers use different rates of increase of the radial distance, but, as we will see, their choices for angles of displacement seem to be much more limited.

The angle α in Figure 1.4.6 is chosen to be 90°. Real flowers do not like angles of displacement that go evenly into the full circle of 360°. Their reason is very simple and can be summarized in one word: efficiency. Figure 1.4.7 shows a cyber-flower with a displacement angle of $\alpha = 90°$. We observe what goes wrong: the seeds pile at rays 90° apart.

Let us change the angle of displacement slightly, to 91° (see Figure 1.4.8). There is a significant change: we see some spirals. But this is still far from resembling a real sunflower. So then, the natural question to ask is "what are the angles of displacement that 'real' sunflowers use?"

It turns out that sunflowers have an almost exclusive preference for displacement angles obtained by subdividing 360° by the golden ratio ϕ or powers of the golden ratio. Then do we get the *right*, healthy looking flowers, and only then do we have the two numbers of clockwise and counterclockwise spirals being two consecutive Fibonacci numbers.

Here, in Figures 1.4.9 and 1.4.10, are two of them. The angle of displacement α for the first cyber-flower is $360°/\phi$, whereas for the second one it is $360°/\phi^2$.

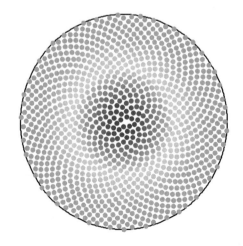

FIGURE 1.4.9 $\alpha = 360°/\phi$.

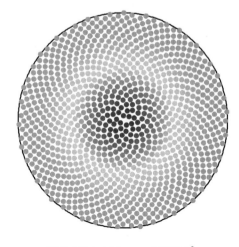

FIGURE 1.4.10 $\alpha = 360°/\phi^2$.

Exercise: How many clockwise, and how many counterclockwise spirals appear in each of the cyber-flowers shown in Figures 1.4.9 and 1.4.10? Do we get two consecutive Fibonacci numbers?

In Figure 1.4.11 we show a cyber-flower for which the radial distance of the nth seed is n units, and such that the radial angle of displacement is $\alpha = 360°/\phi$. In Figure 1.4.12 we change the radial distance of the nth seed to 1.01^n units, and we keep the same radial angle of displacement. The numbers of clockwise and counterclockwise spirals are again two consecutive Fibonacci numbers.

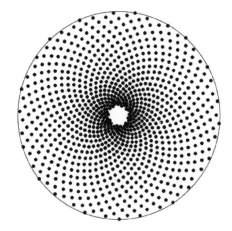

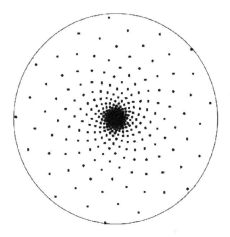

FIGURE 1.4.11 $\alpha = 360°/\phi$ and the distance from the nth seed to the center is n units.

FIGURE 1.4.12 $\alpha = 360°/\phi$ and the distance from the nth seed to the center is 1.01^n units.

We will encounter the cyber-flower in Figure 1.4.12 again when we talk about similarities (Section 3.1) and fractals (Section 3.3).

How the Fibonacci Numbers are Linked to the Golden Ratio

Here is an explicit formula for the nth Fibonacci number f_n : $f_n = \dfrac{(1 + \sqrt{5})^n - (1 - \sqrt{5})^n}{2^n \sqrt{5}}$.

This is called the Binet formula (even though it was known at least 100 years before Binet, who lived in the nineteenth century). We will neither derive* it nor use it again in this textbook. It is displayed here only so that we can observe two interesting features.

Amazingly, even though the formula involves the irrational number[†] $\sqrt{5}$ it does simplify to the Fibonacci number f_n, which for each n is a positive integer.

Secondly, recalling that the golden ratio ϕ is exactly $\dfrac{1+\sqrt{5}}{2}$, it seems that something similar to that expression lurks behind the above formula. Indeed, it turns out (and it is

---

* One way to get this formula is by utilizing eigenvectors, a subject of a standard intermediate linear algebra course.

[†] A number is irrational if it cannot be written as n/m for some integers n and m.

not hard to check) that $f_n = \dfrac{\phi^n - (1/\phi)^n}{\phi + \dfrac{1}{\phi}}$. We can now see how Fibonacci numbers are linked to the golden ratio.

The link between these two goes even deeper. Here is what happens with the quotients of the initial consecutive Fibonacci numbers.

$$\frac{f_2}{f_1} = \frac{1}{1} = 1, \frac{f_3}{f_2} = \frac{2}{1} = 2, \frac{f_4}{f_3} = \frac{3}{2} = 1.5, \frac{f_5}{f_4} = \frac{5}{3} = 1.666 \ldots, \frac{f_6}{f_5} = \frac{8}{5} = 1.6, \frac{f_7}{f_6} = \frac{13}{8} = 1.625,$$

$$\frac{f_8}{f_7} = \frac{21}{13} = 1.615 \ldots, \frac{f_9}{f_8} = \frac{34}{21} = 1.619 \ldots, \frac{f_{10}}{f_9} = \frac{55}{34} = 1.617 \ldots, \frac{f_{11}}{f_{10}} = \frac{89}{55} = 1.618 \ldots, \text{where}$$

ellipses "..." are used when the decimals do not end. We notice that $\dfrac{f_{11}}{f_{10}} = 1.618\ldots$, matching the golden ratio at the first three decimal places. This is not a coincidence: as n becomes larger and larger, the quotient $\dfrac{f_{n+1}}{f_n}$ becomes closer and closer to the golden ratio ϕ. Using a more sophisticated terminology, we can say that the limit of the quotient $\dfrac{f_{n+1}}{f_n}$, as n goes to infinity, is equal to ϕ. A step-by-step partial justification of this claim is deferred to Exercise 7.

Exercises:

1. Count the number of clockwise and counterclockwise spirals in the following flowers.

2. Compute f_{12}.
3. a. The 27th Fibonacci number is 196,418, and the 26th Fibonacci number is 121,393. Find the 24th Fibonacci number.
   b. We know that f_{19} is 4181 and f_{22} is 17711. What is f_{20} and what is f_{21}? [*Note*: there is a shorter solution than the slow walk from f_1 all the way to f_{20} and f_{21}.]
4. Define a sequence of "gibonacci" numbers g_n as follows: $g_1 = 3$, $g_2 = 7$, $g_n = g_{n-1} + g_{n-2}$, $n = 3, 4, \ldots$ (that is, starting from g_3 each of these numbers is the sum of the previous two).
   a. Write down the first six terms of the sequence of gibonacci numbers.
   b. It can be shown that $g_{14} = 2063$ and that $g_{16} = 5401$. Compute g_{15}.
5. (Adapted from the book *The Golden Section* by Hans Walser) We construct a sequence of rectangles by adding squares over the larger sides of the rectangles (see the picture; assume that the length of the side of the initial square is 1 unit).

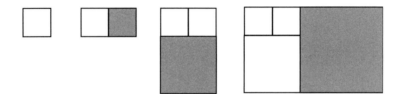

a. Draw the next rectangle in this sequence.

b. Explain how the shorter sides of the rectangles in this sequence are related to the Fibonacci numbers.

6. Each black kangaroo bears one white and one black baby kangaroo during its lifespan. Each white kangaroo bears only one black baby kangaroo in its life. We start with one white and one black kangaroo in the first generation.

a. How many kangaroos are in the fourth generation?

b. Denote by k_n the number of kangaroos in generation number n. What is the relationship between the numbers $k_1, k_2, ..., k_n, ...$ and the Fibonacci numbers?

7. Let us assume that the sequence of fractions $\dfrac{f_2}{f_1}, \dfrac{f_3}{f_2}, \dfrac{f_4}{f_3}, \dfrac{f_5}{f_4}, ..., \dfrac{f_{n+1}}{f_n}, ...$ approaches some number x.

a. Convince yourself that the sequence of fractions $\dfrac{f_3}{f_2}, \dfrac{f_4}{f_3}, \dfrac{f_5}{f_4}, ..., \dfrac{f_{n+2}}{f_{n+1}}, ...$ (the first fraction from the original sequence is omitted) also approaches x.

b. Show that $\dfrac{f_{n+2}}{f_{n+1}} = 1 + \dfrac{f_n}{f_{n+1}}$.

c. Consider the sequence of numbers $1 + \dfrac{f_n}{f_{n+1}}$ that we get as n changes through the set 1, 2, 3, ... of positive integers. Use the assumption that the sequence $\left[\dfrac{f_2}{f_1}, \dfrac{f_3}{f_2}, \dfrac{f_4}{f_3}, \dfrac{f_5}{f_4}, ..., \dfrac{f_{n+1}}{f_n}, ...\right]$ approaches to show that this sequence of numbers approaches $1 + \dfrac{1}{x}$.

d. Conclude from (a), (b), and (c) that $x = 1 + \dfrac{1}{x}$.

e. Solve $x = 1 + \dfrac{1}{x}$ and confirm that x is the golden ratio ϕ.

CHAPTER 2

Plane Transformations

A planar symmetry is a *rigid transformation* (rearrangement) of the points in the plane. In this chapter, we consider planar symmetries and some other related geometrical notions (e.g., wallpaper designs).

2.1 PLANE SYMMETRIES

We fix one plane throughout, and we assume that all of the two-dimensional objects we encounter in this chapter are in that fixed plane.

A **transformation** of the points in the plane is a rearrangement of all the points in the plane. If no two points are moved into a single position, then we say that the transformation is **one-to-one**. A transformation is **onto** if all of the positions in the plane are achieved by some points in the rearrangement. A **bijection** is a transformation that is both onto and one-to-one. All of the transformations we encounter in this section are bijections.

We now consider *rigid* transformations of the points of that plane. A transformation is **rigid** if it preserves distances. Such transformations are also called **symmetries** or *isometries*. We exclusively call them symmetries. Symmetries are always bijections (Exercise 9).

Denote a transformation of the (points of the) plane by *f*. Given a point *A* in the plane, we denote the point we get after we apply the transformation *f* to the point *A* by *f(A)*. You can think of *f* as a particular rule that tells us where any point on the plane is being moved. We point out that it is where points are moved that will be of importance to us, and that *that* only determines a transformation, not how they are moved.

In the following examples, we illustrate various symmetries. In these illustrations we show what happens only to a few chosen points after the action of a symmetry. However, the reader should bear in mind that every symmetry acts on every point in the infinite, unbounded plane.

Example 1. A Rotation

Consider the following simple example. Fix a point O in the plane and let f be the **rotation** around O (the **center of rotation**) through 45° (the **angle of rotation**). Keep in mind that all of our objects are planar (so, Dadat in Figure 2.1.1 should be viewed as being two-dimensional).

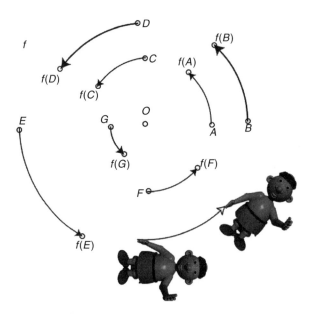

FIGURE 2.1.1 Every point in the plane is rotated around O through 45° by the rotation denoted by f.

We see how each point is rotated around O. Since O is not moved anywhere we have that $f(O) = O$; we say that the point O is fixed by f. Every rotation in the plane is uniquely specified by the center of rotation and by the angle of rotation. □

Using the notion we have introduced, we may describe rigid transformations f as follows: For every two points A and B in the plane, the distance between A and B is the same as the distance between the image points $f(A)$ and $f(B)$. Every rotation preserves distances, and hence it is a symmetry.

We describe below two other types of symmetries of the plane.

Example 2. A Translation

In Figure 2.1.2, we depict a **translation** (denoted by h). Every point in the plane is moved along a fixed vector, called the **vector of translation**. In general, a **vector** is another word for an oriented arrow. Every translation is uniquely determined by its vector of translation. □

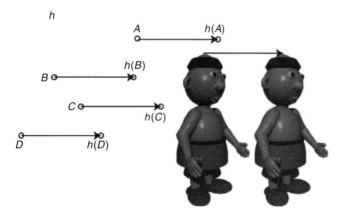

FIGURE 2.1.2 The translation h moves every point in the plane in the direction of the given vector.

□

Example 3. A Reflection

Figure 2.1.3 is an illustration of a **reflection** with respect to a fixed line, called *the* **line of reflection**. We can think of the line as a two-sided mirror: every point on one side of the line is reflected into a point on the other side, and vice versa. So, in the illustration we have that $g(A) = B$, but also that $g(B) = A$. The points on the line l are not moved. Every reflection is uniquely determined by its line of reflection.

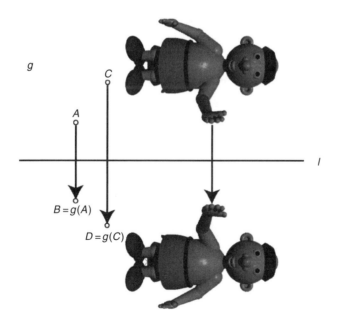

FIGURE 2.1.3 This figure shows what the reflection g does to the points in the plane.

Since the line l, the line of reflection, acts as a two-sided mirror, both copies of Dadat are reflections of each other.

□

We back up a bit to note a couple of conventions. Rotations by positive angles traditionally denote rotations in the counterclockwise direction; rotations through negative angles are clockwise. We will adopt that convention. For example, a rotation through −60° is in the clockwise direction. We should be careful: since rotation through −60° rearranges the points on the plane the same way as the rotation by the positive angle of 300°, we consider these two to be equal even though we associate different movements of the points with these two rotations. We emphasize this once again.

> Two transformations f and g are considered to be equal if $f(A) = g(A)$ for every point A in the plane.

The three symmetries we have encountered so far (a rotation, a translation, and a reflection) are not the only types of symmetries. For example, if we first apply a rotation and follow it by a translation, we get a new symmetry. The symmetry we get as a result of repeatedly applying symmetries is called a *composition* of symmetries. In general, a **composition** of a transformation f followed by a transformation g, denoted $g \circ f$, is the transformation of the points of the plane obtained by first applying f and then applying g. So, for every point A in the plane, we have $g \circ f(A) = g(f(A))$. Alternatively, the image of the point A under the composition $g \circ f$ is obtained by first applying f to the point A to get $f(A)$, and then applying g to the point $f(A)$ to get the final image $g(f(A))$. Note the somewhat unusual order of the transformations in our notation: in the composition $g \circ f$, we first apply the transformation f. That order is fairly standard and it is convenient since when we write $g \circ f(A)$ for the image of the point A, the transformation that acts first (f) is closer to the point A.

Example 4. A Composition of Two Symmetries

In Figure 2.1.4 we depict the composition of a rotation by 45°, followed by a translation in the direction of the indicated vector.

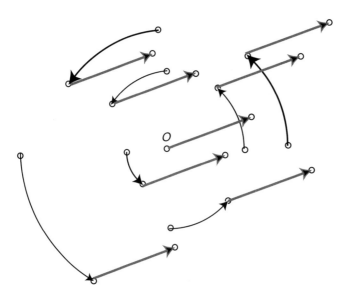

FIGURE 2.1.4 Every point on the plane (just a few of them are explicitly shown here) is first rotated about the point O through 45°, and then translated by the vector indicated in the picture. □

Exercise: Suppose t_1 is a translation along a vector of unit length in the upward direction and suppose t_2 is the translation along a unit vector pointing rightward. Describe the symmetry obtained by composing t_1 followed by t_2.

The most fundamental problem related to plane symmetries is the *classification problem*: can we describe all the symmetries in the plane? What kinds of symmetries are there? It turns out that there is a surprisingly simple description of the planar symmetries. We state it as a separate theorem, and then we give a sketch of the proof. □

Theorem: (The Classification Theorem for Plane Symmetries) Every symmetry of the plane is either a composition of a translation followed by a rotation, or it is a composition of a translation followed by a reflection.

A Bit of Math. An Outline of a Proof of the Classification Theorem

We will give an outline of the proof of this theorem. Suppose *f* is any symmetry. Then it sends the vertices of $\triangle ABC$ to the vertices of $\triangle f(A)f(B)f(C)$ (see Figures 2.1.5 and 2.1.6). Since *f* preserves distances, these two triangles are congruent, and so their corresponding sides are of equal length, for example, $AB = f(A)f(B)$. The symmetry *f* is the only symmetry that sends the vertices of $\triangle ABC$ to the vertices of $\triangle f(A)f(B)f(C)$. That is true since *f* preserves distances, so that the distances between any other point *D* and the points *A*, *B*, and *C* are, respectively, equal to the distances between its image *f(D)* and the points *f(A)*, *f(B)*, and *f(C)*. Since there is exactly one point at given distances from *f(A)*, *f(B)*, and *f(C)*, the point *f(D)* is uniquely determined by the positions of the points *f(A)*, *f(B)*, and *f(C)*.

In the next part of the argument, we find a composition of two basic symmetries as indicated in the statement of the theorem that sends the vertices of $\triangle ABC$ to the vertices of $\triangle f(A)f(B)f(C)$; it would then follow from the preceding two sentences that that composition must be *f*.

We first translate through the vector extending from *A* to *f(A)*. Obviously, that would move *A* to *f(A)*, and the points *B* and *C* to some points *B'* and *C'*, respectively (see Figures 2.1.5 and 2.1.6). Since translations preserve distance, $\triangle f(A)B'C'$ is also congruent to $\triangle ABC$. So, it is also congruent to $\triangle f(A)f(B)f(C)$. There are now two possibilities, depending on the orientation of $\triangle f(A)B'C'$ (as indicated in the captions to Figures 2.1.5 and 2.1.6). □

A few more remarks are in order. First of all, there exists a special translation, and a special rotation: the translation through the vector of length zero (the **zero vector**), and the rotation through the zero angle, respectively. In both cases, we do not move any of the points of the plane. The symmetry that lets all points stay at their original positions is called the **identity symmetry**. The name comes from the following easy property: the composition of any symmetry *f* followed by the identity symmetry is the symmetry *f* itself. In the classification theorem, we allow the translations and rotations to be equal to the identity.

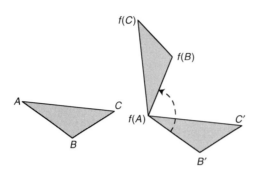
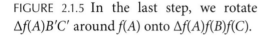

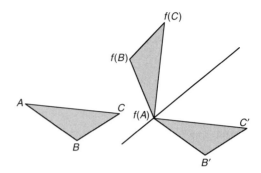

FIGURE 2.1.5 In the last step, we rotate $\Delta f(A)B'C'$ around $f(A)$ onto $\Delta f(A)f(B)f(C)$.

FIGURE 2.1.6 However, if the triangles are of opposite orientation, then we reflect $\Delta f(A)B'C'$ with respect to the indicated line to $\Delta f(A)f(B)f(C)$.

Consequently, the classification does include the three basic symmetries (rotation = rotation followed by the identity translation, translation = identity rotation followed by a translation, and reflection = reflection followed by the identity translation). ☐

Example 5. Glide Reflections

Reflections followed by translations in the direction parallel to the line of reflection are usually called **_glide reflections_**.

A glide reflection is illustrated in Figure 2.1.7, where first we apply the reflection r to each and every point in the plane, and then we follow it by the translation t of all of the

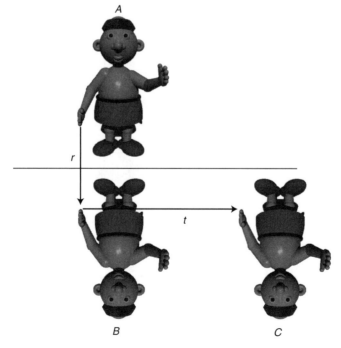

FIGURE 2.1.7

points in the plane. In that figure, we only show what happens with the (two-dimensional) copy A of Dadat: first we reflect it to get the intermediate position B, and then we translate it to its final destination C. The translation vector is parallel to the reflection line, as it should be in all glide reflections. □

Example 6. Constructing the Center and the Angle of a Rotation

The flat character shown at the bottom of Figure 2.1.8 is obtained from the other by rotating it around a point O and through a negative angle of rotation. Construct (as always, using an unmarked ruler and a compass) the center of rotation and the angle of rotation.

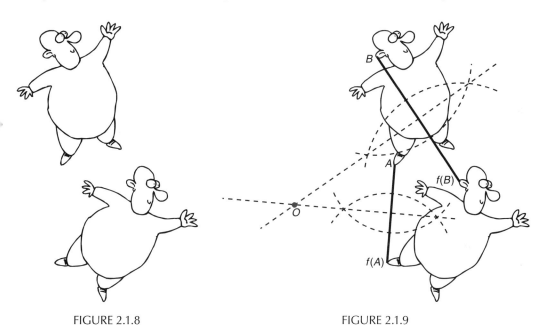

FIGURE 2.1.8   FIGURE 2.1.9

Solution: Denote the rotation by f. For every point A, the pair of points A and $f(A)$ lie on a circle centered at the center of rotation. That center must be somewhere on the line bisecting the chord from A to $f(A)$. We identify two such pairs of points in Figure 2.1.9 (A and $f(A)$, then B and $f(B)$). The center of rotation is at the intersection O of the bisectors of the chords $Af(A)$ and $Bf(A)$ (as shown in Figure 2.1.9). It does not matter which two pairs we choose, but it is important that the two points in any such pair are obtained by rotating one of them into the other. The angle of rotation is the angle $AOf(A)$ in that direction (clockwise, since the angle was given to be negative). □

Example 7. Constructing the Line of Reflection

The two two-dimensional characters in Figure 2.1.10 are obtained by a reflection. Construct the line of reflection.

Solution: Choose any pair of points in the plane such that they are mutual images under the reflection (we choose two corresponding heels of the two characters in Figure 2.1.11).

FIGURE 2.1.10

FIGURE 2.1.11

Construct (using a ruler and a compass) the bisector of the line segment between the two chosen points. That bisector is the line of reflection we sought. □

Exercises: In the following problems the word "construct" is short for "construct using an unmarked ruler and a compass."

1. The triangle *B* (in Figure 2.1.12) is obtained from the triangle *A* by applying a symmetry to the points on the plane. Identify that symmetry.

2. Choose points *P* and *O* on the plane and choose a line *l* not passing through any of these two points. Construct the image of the point *P* after applying to it a rotation of 60° centered at *O*, followed by a reflection with respect to the line *l*.

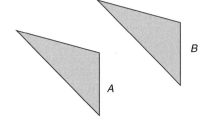

FIGURE 2.1.12

3. Suppose *f* is a translation through a fixed vector. What is the composition of *f* followed by *f*?

4. a. Suppose *l* and *m* are two parallel lines. What is the composition of the reflection with respect to *l* followed by the reflection with respect to *m*?

   b. Suppose *l* and *m* are two intersecting lines. What is the composition of the reflection with respect to *l* followed by the reflection with respect to *m* in this case?

5. Suppose *h* is a rotation around a point *O* through 125°, and suppose *g* is a rotation around the same point *O* through −75°. Describe the composition of *h* followed by *g*. Is it the same as the composition of *g* followed by *f*?

6. Consider the reflection with respect to a fixed point *O*—that is, every point *A* is sent to the point *B* on the other side of *O* and such that the segments *OA* and *OB* are equal. Show that this reflection is the same as the rotation centered at *O* and through 180°.

7. The triangle *B* in Figure 2.1.13 is obtained from the triangle A by applying a symmetry to the points on the plane. Describe that symmetry. [*Hint*: a composition of two symmetries.]

8. There are six symmetries of the plane that send every point of the design shown in Figure 2.1.14 to a point within the same design. Find them.

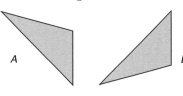

FIGURE 2.1.13

9. a. Show that every symmetry is one-to-one. [*Hint*: this is easy.]
   b. Show that every symmetry is a bijection.
10*. Show that symmetries preserve angles. That is, show that if *f* is a symmetry and *A*, *O*, and *B* are three points on the plane, then the angle *AOB* is the same as the angle *f(A)f(O)f(B)*.
11. Show that the composition of a rotation around a point *O* by an angle α followed by a translation through a vector *v* is the same as the composition of a translation through the vector *v* followed by the rotation through the same angle α around the point *O'* obtained by translating *O* through *v*. [*Hint*: it suffices to show that these two compositions agree at three points that are not on the same line.]

FIGURE 2.1.14

FIGURE 2.1.15 A moth: one reflectional symmetry.

12. a. The moth in Figure 2.1.15 is symmetric with respect to a reflection. Identify the line of that reflection.
   b. The (unbounded) curve *l* shown in Figure 2.1.17 divides the plane into two parts, and each vertical line intersects it at a single point. We define a transformation *f* as follows: for each point X_1 above the curve *l* the image $f(X_1)$ is the point vertically below X_1, on the other side of *l*, and at the same vertical distance to *l*. Similarly, for each point X_2 below *l*, the image $f(X_2)$ is the point vertically above X_2, on the other side of *l* and at the same vertical distance to *l*. In Figure 2.1.17 we see that the points A_1, B_1, C_1 which are sent via *f* to the points A_2, B_2, C_2 and vice versa. Explain why this transformation is not a symmetry. [Figure 2.1.16 shows an illustration of the transformation *f* in this problem.]

FIGURE 2.1.16 A reflection with respect to a curve.

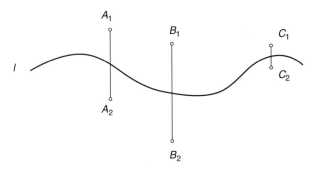

FIGURE 2.1.17

2.2* PLANE SYMMETRIES, VECTORS, AND MATRICES (OPTIONAL)

First we recall the Descartes coordinate system: it consists of a pair of perpendicular, oriented lines (say, one horizontal and the other vertical) graded with a unit length. Given any such coordinate system, we can associate a pair of numbers to any point in the plane: the first number (the first coordinate) is the signed distance to the vertical line, and the second number (the second coordinate) is the signed distance to the horizontal line. We say *signed distance* to indicate that these numbers could be positive or negative, depending on whether the point is to the right or to the left of the vertical line, respectively (for the first coordinate), and depending on whether the point is above or below the horizontal coordinate line, respectively (for the second coordinate). In Figure 2.2.1, we see a coordinate system in the plane and a few points with their coordinates.

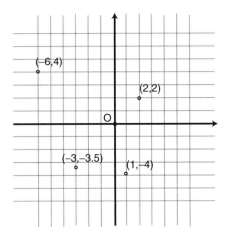

FIGURE 2.2.1 A Descartes coordinate system: each point in the plane corresponds to a pair of numbers.

The intersection point of the two coordinate lines is the **origin** and it has coordinates $(0, 0)$. The coordinate lines are called the **axes** of the coordinate system.

The association "point ⇔ pair of numbers" goes both ways. Because of that, this correspondence allows us to encode every planar object with the set of pairs of numbers corresponding to the points of that object. For example, the horizontal coordinate line (often called the *x*-axis) corresponds to the set of all pairs of numbers with the second coordinate equal to 0.

The Distance Formula

There are many advantages of the above correspondence between points and pairs of numbers. One of them is that knowing the coordinates of the points of an object allows us to analyze that object analytically, that is, through the numbers in the coordinates of its points. For example, it follows from the Pythagorean theorem that the distance between two points (a_1, a_2) and (b_1, b_2) is $\sqrt{(a_1 - b_1)^2 + (a_2 - b_2)^2}$. A justification is provided in Figure 2.2.2.

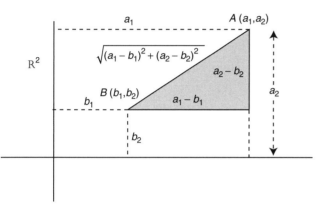

FIGURE 2.2.2 The distance formula for the points in the plane given through their coordinates.

Our main goal in this section is to describe the action of the basic symmetries in terms of the coordinates of the points. For example, is there a simple procedure to find the coordinates of the point $(1, -4)$ shown in Figure 2.2.1 after we rotate it by 45° around the origin? Is there such a procedure in the cases of rotations by any angle and around any point? What about a procedure that will give the images of points moved by translations or reflections? We will answer these questions affirmatively, and construct a small, elegant machine that will do all of these procedures for us.

An immediate advantage of our correspondence "point ⇔ pair of numbers" is that we can now perform algebraic operations on points through their coordinates.

For convenience (to make the operations that we are about to define outwardly more elegant), we sometimes write the pairs of numbers in columns. Thus, for example, instead of $(1, 2)$ we will write $\begin{bmatrix} 1 \\ 2 \end{bmatrix}$, and it should be understood that the top number is the first coordinate of the corresponding point, and the bottom number is the second coordinate. With that notation, the following definition of addition of points (or pairs of numbers) is virtually forced upon us.

Given any two pairs of numbers (or points) $\begin{bmatrix} a_1 \\ a_2 \end{bmatrix}$ and $\begin{bmatrix} b_1 \\ b_2 \end{bmatrix}$, their *sum*, denoted $\begin{bmatrix} a_1 \\ a_2 \end{bmatrix} + \begin{bmatrix} b_1 \\ b_2 \end{bmatrix}$ is defined to be the pair of numbers $\begin{bmatrix} a_1 + b_1 \\ a_2 + b_2 \end{bmatrix}$.

For example, $\begin{bmatrix} 1 \\ 2 \end{bmatrix} + \begin{bmatrix} 3 \\ 4 \end{bmatrix} = \begin{bmatrix} 1 + 3 \\ 2 + 4 \end{bmatrix} = \begin{bmatrix} 4 \\ 6 \end{bmatrix}$ and $\begin{bmatrix} -1 \\ 0.5 \end{bmatrix} + \begin{bmatrix} -0.2 \\ 33 \end{bmatrix} = \begin{bmatrix} -1 + (-0.2) \\ 0.5 + 33 \end{bmatrix} = \begin{bmatrix} -1.2 \\ 33.5 \end{bmatrix}$.

With this addition of points we are ready to algebraically encode translations; that is, we can now describe them in terms of additions of pairs of numbers.

Example 1. Translations

Fix one point (one pair of numbers), say, $\begin{bmatrix} 1 \\ 2 \end{bmatrix}$. We will first compute a few sums of that point with some other points in the coordinate plane:

$$\begin{bmatrix} 1 \\ 2 \end{bmatrix} + \begin{bmatrix} -1 \\ 1 \end{bmatrix} = \begin{bmatrix} 0 \\ 3 \end{bmatrix}$$

$$\begin{bmatrix} 1 \\ 2 \end{bmatrix} + \begin{bmatrix} 0 \\ 0 \end{bmatrix} = \begin{bmatrix} 1 \\ 2 \end{bmatrix}$$

$$\begin{bmatrix} 1 \\ 2 \end{bmatrix} + \begin{bmatrix} -2 \\ -3 \end{bmatrix} = \begin{bmatrix} -1 \\ -1 \end{bmatrix}$$

$$\begin{bmatrix} 1 \\ 2 \end{bmatrix} + \begin{bmatrix} 1 \\ 2 \end{bmatrix} = \begin{bmatrix} 2 \\ 4 \end{bmatrix}$$

$$\begin{bmatrix} 1 \\ 2 \end{bmatrix} + \begin{bmatrix} -2 \\ 2 \end{bmatrix} = \begin{bmatrix} -1 \\ 4 \end{bmatrix}$$

$$\begin{bmatrix} 1 \\ 2 \end{bmatrix} + \begin{bmatrix} 4 \\ -3 \end{bmatrix} = \begin{bmatrix} 5 \\ -1 \end{bmatrix}$$

We draw an obvious conclusion that if we add $\begin{bmatrix} 1 \\ 2 \end{bmatrix}$ to a point in a plane (written as a pair of numbers), the result is another point in the plane. So, addition with $\begin{bmatrix} 1 \\ 2 \end{bmatrix}$ rearranges the points in the plane, each one moving to its sum with $\begin{bmatrix} 1 \\ 2 \end{bmatrix}$. For example, we have computed above that $\begin{bmatrix} -1 \\ 1 \end{bmatrix}$ is moved to $\begin{bmatrix} 0 \\ 3 \end{bmatrix}$, $\begin{bmatrix} 0 \\ 0 \end{bmatrix}$ is moved to $\begin{bmatrix} 1 \\ 2 \end{bmatrix}$, $\begin{bmatrix} -2 \\ -3 \end{bmatrix}$ is moved to $\begin{bmatrix} -1 \\ -1 \end{bmatrix}$, $\begin{bmatrix} 1 \\ 2 \end{bmatrix}$ is moved to $\begin{bmatrix} 2 \\ 4 \end{bmatrix}$, $\begin{bmatrix} -2 \\ 2 \end{bmatrix}$ is moved to $\begin{bmatrix} -1 \\ 4 \end{bmatrix}$, and $\begin{bmatrix} 4 \\ -3 \end{bmatrix}$ is moved to $\begin{bmatrix} 5 \\ -1 \end{bmatrix}$. What kind of rearrangement do we get? The answer will be obvious once we visualize our computation (Figure 2.2.3).

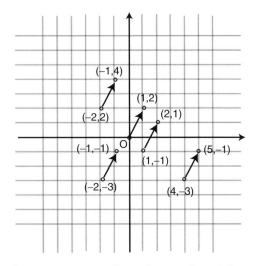

FIGURE 2.2.3 We plot a few points, as well as the results of the addition of these points with (1,2).

As Figure 2.2.3 suggests, the addition of the points in the plane by our fixed point effects a translation of the points in the plane by a fixed vector. What is that fixed vector? That is also easy to see: when we add any point to the fixed point $\begin{bmatrix} 1 \\ 2 \end{bmatrix}$, the first coordinate of the former increases by 1, while the second increases by 2. So, the fixed vector of the translation that results from the addition by $\begin{bmatrix} 1 \\ 2 \end{bmatrix}$ is oriented in such a way that it is 1 unit to the right and 2 units upward from any given point. If we fix the starting point of the vector to be the origin, then its end point is the point $\begin{bmatrix} 1 \\ 2 \end{bmatrix}$. The vector that starts at the origin and ends at the point A in the plane is called the **position vector** of the point A. Our final conclusion about this example is that addition by $\begin{bmatrix} 1 \\ 2 \end{bmatrix}$ is the same as translation by the position vector of the point $\begin{bmatrix} 1 \\ 2 \end{bmatrix}$.

Only a very small modification is needed to get the following more general conclusion.

> (**Algebraic description of translations**) Addition of all the points in the plane by a fixed point $\begin{bmatrix} a \\ b \end{bmatrix}$ results in the translations of the points of the plane by the position vector of the point $\begin{bmatrix} a \\ b \end{bmatrix}$.

So, if we are given a translation through a vector v, to describe it algebraically, it suffices to find the point A_v in the plane such that its position vector is of the same orientation and size as the vector v. That, in turn, can be achieved by positioning the vector v so that it starts at the origin; the end point will then be A_v.

For example, here is how we can encode algebraically the translation through the vector v pointing 2 units to the right and 1 unit downward from its initial point. First, we position v so that it starts at the origin: a small sketch (and bearing in mind where v points) will easily convince you that its terminal vertex would be the point $\begin{bmatrix} 2 \\ -1 \end{bmatrix}$. By our analysis above, to get the image of any point $\begin{bmatrix} a \\ b \end{bmatrix}$ under trans$_v$, we should simply add $\begin{bmatrix} 2 \\ -1 \end{bmatrix} + \begin{bmatrix} a \\ b \end{bmatrix}$. □

The other two basic symmetries are only slightly more complicated to encode algebraically. In order to do that, we first need to introduce *matrices*.

Matrices

A **two-by-two matrix** $\begin{bmatrix} a_1 & a_2 \\ a_3 & a_4 \end{bmatrix}$ is a square array of four numbers a_1, a_2, a_3, and a_4. In other words, it is simply a quadruple of numbers written in a special way. The numbers a_1, a_2, a_3, and a_4 in the matrix are called the **entries** of the matrix. In general, an **n-by-m matrix** is a rectangular array of numbers with n rows and m columns. The numbers n and m in this setting are called the **dimensions** of the matrix.

Since we will mostly deal with two-by-two matrices in this section, we will often drop that attribute and simply use the word *matrix*. For example, $\begin{bmatrix} 1 & 2 \\ -2 & 1 \end{bmatrix}, \begin{bmatrix} 0.2 & 22 \\ \pi & \phi \end{bmatrix}, \begin{bmatrix} 1 & 0 \\ 0 & 1 \end{bmatrix}$, and $\begin{bmatrix} 0 & 0 \\ 0 & 0 \end{bmatrix}$ are all matrices.

Some of the algebraic operations that can be defined over matrices are simple enough. Both addition of matrices and multiplication of a matrix by a number are defined in the expected way—entrywise. For example, here is how we add the matrices $\begin{bmatrix} 1 & 2 \\ -2 & 1 \end{bmatrix}$ and $\begin{bmatrix} 3 & 4 \\ 5 & 6 \end{bmatrix}$:

$$\begin{bmatrix} 1 & 2 \\ -2 & 1 \end{bmatrix} + \begin{bmatrix} 3 & 4 \\ 5 & 6 \end{bmatrix} = \begin{bmatrix} 1+3 & 2+4 \\ -2+5 & 1+6 \end{bmatrix} = \begin{bmatrix} 4 & 6 \\ 3 & 7 \end{bmatrix}$$

and here we multiply the matrix $\begin{bmatrix} 1 & 2 \\ -2 & 1 \end{bmatrix}$ by the number 3:

$$3\begin{bmatrix} 1 & 2 \\ -2 & 1 \end{bmatrix} = \begin{bmatrix} (3)(1) & (3)(2) \\ (3)(-2) & (3)(1) \end{bmatrix} = \begin{bmatrix} 3 & 6 \\ -6 & 3 \end{bmatrix}$$

We will now define a strange but important operation called multiplication of a matrix and a point. Start with any matrix $\begin{bmatrix} a_1 & a_2 \\ b_1 & b_2 \end{bmatrix}$ and any pair of numbers $\begin{bmatrix} c_1 \\ c_2 \end{bmatrix}$* (we will again prefer to write pairs of numbers in columns). The **product** of the matrix $\begin{bmatrix} a_1 & a_2 \\ b_1 & b_2 \end{bmatrix}$ and the pair $\begin{bmatrix} c_1 \\ c_2 \end{bmatrix}$ (in that order), denoted $\begin{bmatrix} a_1 & a_2 \\ b_1 & b_2 \end{bmatrix}\begin{bmatrix} c_1 \\ c_2 \end{bmatrix}$, is defined by the following formula:

(The Product formula) $\quad \begin{bmatrix} a_1 & a_2 \\ b_1 & b_2 \end{bmatrix}\begin{bmatrix} c_1 \\ c_2 \end{bmatrix} = \begin{bmatrix} a_1 c_1 + a_2 c_2 \\ b_1 c_1 + b_2 c_2 \end{bmatrix}.$

This may look somewhat ugly. For the time being, it is important just to accept and trust that, as ugly as it is, this multiplication will be the core of the theory that we are developing. □

Example 2. Multiplication by a Matrix

We will find the product of $\begin{bmatrix} 1 & 2 \\ 3 & 4 \end{bmatrix}$ and $\begin{bmatrix} -1 \\ 0 \end{bmatrix}$ (notice that, according to the product formula, we should multiply the numbers of the first row of the matrix with the coordinates of the point, and then repeating the same procedure for the second row of the matrix):

$$\begin{bmatrix} 1 & 2 \\ 3 & 4 \end{bmatrix}\begin{bmatrix} -1 \\ 0 \end{bmatrix} = \begin{bmatrix} 1 \cdot (-1) + 2 \cdot 0 \\ 3 \cdot (-1) + 4 \cdot 0 \end{bmatrix} = \begin{bmatrix} -1+0 \\ -3+0 \end{bmatrix} = \begin{bmatrix} -1 \\ -3 \end{bmatrix}$$

Observe that the result is again a pair of numbers. It follows from our product formula that we always end with a pair of numbers (written in a column). So, when we multiply the matrix $\begin{bmatrix} 1 & 2 \\ 3 & 4 \end{bmatrix}$ by the pair $\begin{bmatrix} -1 \\ 0 \end{bmatrix}$ we get the pair $\begin{bmatrix} -1 \\ -3 \end{bmatrix}$. If we view pairs of numbers as points in the plane with a coordinate system (as established at the beginning of this section), then we can visualize and associate movements of points to multiplication by the matrix $\begin{bmatrix} 1 & 2 \\ 3 & 4 \end{bmatrix}$. For example, we saw that the point with coordinates $(-1, 0)$ $\left(\text{or } \begin{bmatrix} -1 \\ 0 \end{bmatrix}\right)$ is sent to the point $(-1, -3)$ $\left(\text{or } \begin{bmatrix} -1 \\ -3 \end{bmatrix}\right)$ after we multiply the former by the matrix $\begin{bmatrix} 1 & 2 \\ 3 & 4 \end{bmatrix}$.

---

* The pairs of numbers $\begin{bmatrix} c_1 \\ c_2 \end{bmatrix}$ can also be viewed as matrices, specifically as 2×1 matrices.

Let us apply multiplication by the same matrix again, this time to the point $\begin{bmatrix}1\\0\end{bmatrix}$: $\begin{bmatrix}1&2\\3&4\end{bmatrix}\begin{bmatrix}1\\0\end{bmatrix} = \begin{bmatrix}1\cdot1+2\cdot0\\3\cdot1+4\cdot0\end{bmatrix} = \begin{bmatrix}1\\3\end{bmatrix}$. So we can say that the point $\begin{bmatrix}1\\0\end{bmatrix}$ is moved to the point $\begin{bmatrix}1\\3\end{bmatrix}$ when we multiply the former by $\begin{bmatrix}1&2\\3&4\end{bmatrix}$. Finally, let us multiply $\begin{bmatrix}1&2\\3&4\end{bmatrix}$ by $\begin{bmatrix}-2\\2\end{bmatrix}$: $\begin{bmatrix}1&2\\3&4\end{bmatrix}\begin{bmatrix}-2\\2\end{bmatrix} = \begin{bmatrix}1\cdot(-2)+2\cdot2\\3\cdot(-2)+4\cdot2\end{bmatrix} = \begin{bmatrix}-2+4\\-6+8\end{bmatrix} = \begin{bmatrix}2\\2\end{bmatrix}$. So, multiplication by $\begin{bmatrix}1&2\\3&4\end{bmatrix}$ moves the point $\begin{bmatrix}-2\\2\end{bmatrix}$ to the point $\begin{bmatrix}2\\2\end{bmatrix}$.

We see that multiplication by the matrix $\begin{bmatrix}1&2\\3&4\end{bmatrix}$ rearranges the points in the plane, each one being moved to the product of the matrix $\begin{bmatrix}1&2\\3&4\end{bmatrix}$ with that point (with coordinates written in a column). In Figure 2.2.4, we visualize what we have found so far.

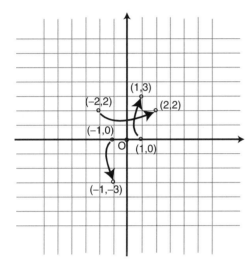

FIGURE 2.2.4 Multiplication by the fixed matrix $\begin{bmatrix}1&2\\3&4\end{bmatrix}$ results in a seemingly complicated transformation of the points of the plane.

We have arrived at the main initial conclusion: multiplication by a fixed matrix results in the rearrangement of the points in the plane. For some matrices these rearrangements will preserve distances (and so, they will result in symmetries), but some of them may result in wilder disturbances of the points in the plane $\left(\text{as was the case with the matrix }\begin{bmatrix}1&2\\3&4\end{bmatrix}\right)$. We are concerned with the former. □

Example 3. Some Reflections

We will now see what kind of rearrangements of the points in the plane results after multiplication by the matrix $\begin{bmatrix}-1&0\\0&1\end{bmatrix}$. This time we go straight to the general case, and check what happens when we multiply the matrix by a point $\begin{bmatrix}a\\b\end{bmatrix}$ with unspecified coordinates a and b: $\begin{bmatrix}-1&0\\0&1\end{bmatrix}\begin{bmatrix}a\\b\end{bmatrix} = \begin{bmatrix}-a\\b\end{bmatrix}$. The relationship between the point $\begin{bmatrix}a\\b\end{bmatrix}$ and the resulting point $\begin{bmatrix}-a\\b\end{bmatrix}$ is obvious: the only difference between the two points is that their first coordinates are mutual negatives. Geometrically that means that both points are at equal distance from the vertical axis, but they are on the opposite sides of it (see Figure 2.2.5).

We conclude that multiplication by $\begin{bmatrix}-1&0\\0&1\end{bmatrix}$ results in a reflection with respect to the vertical axis.

Very similar argument would confirm that multiplication by $\begin{bmatrix} 1 & 0 \\ 0 & -1 \end{bmatrix}$ is the same as reflecting with respect to the horizontal axis. Thus, we have encoded these two reflections in terms of multiplication by matrices.

(**Algebraic description of two reflections**) The reflection with respect to the vertical axis is the same as multiplication by the matrix $\begin{bmatrix} -1 & 0 \\ 0 & 1 \end{bmatrix}$. The reflection with respect to the horizontal axis is the same as multiplication by the matrix $\begin{bmatrix} 1 & 0 \\ 0 & -1 \end{bmatrix}$.

We postpone for a moment the general case of describing reflections with respect to other lines in terms of algebraic operations.

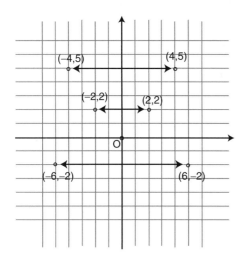

FIGURE 2.2.5 Multiplication by the fixed matrix $\begin{bmatrix} -1 & 0 \\ 0 & 1 \end{bmatrix}$ results in a reflection with respect to the vertical axis. □

Example 4. Rotations about the Origin

This happens to be a somewhat more complicated case. The reader should not be discouraged by the appearance of some basic trigonometric functions. We start with Figure 2.2.6 where we see a point (a, b) rotated through angle α to the point (c, d).

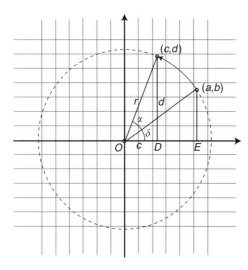

FIGURE 2.2.6 We rotate all points around the origin through the angle α. We show one such point, with general coordinates (a, b).

We would like to express the coordinates of the point (c, d) in terms of a, b, and α. We will focus only on the first coordinate c. The trigonometry of the triangle with vertices at the points O, D, and (c, d) gives immediately that $\cos(\alpha + \delta) = \frac{c}{r}$, from where we easily find $c = r\cos(\alpha + \delta)$. We recall the formula $\cos(\alpha + \delta) = \cos\alpha\cos\delta - \sin\alpha\sin\delta$ and use it to get that $c = r(\cos\alpha\cos\delta - \sin\alpha\sin\delta)$. Expanding this, we get $c = r\cos\alpha\cos\delta - r\sin\alpha\sin\delta$, or, after rearranging a bit for emphasis, $c = (r\cos\delta)\cos\alpha - (r\sin\delta)\sin\alpha$. The trigonometry of the triangle with vertices at O, E, and (a, b) yields that $r\sin\delta = a$ and $r\cos\delta = b$. So, substituting these two, we get $c = a\cos\alpha - b\sin\alpha$.

Very similar argument gives that $d = a\sin\alpha + b\cos\alpha$. So, summarizing what we have found so far, if we rotate the point (a, b) around the origin and by angle α, we get a point with coordinates $(c, d) = (a\cos\alpha - b\sin\alpha, a\sin\alpha + b\cos\alpha)$.

How is this related to our matrices? As it happens, we can write the pair $(a\cos\alpha - b\sin\alpha$, $a\sin\alpha + b\cos\alpha)$ as the result of the product of a relatively simple matrix with the point (a, b): $\begin{bmatrix} a\cos\alpha - b\sin\alpha \\ a\sin\alpha + b\cos\alpha \end{bmatrix} = \begin{bmatrix} \cos\alpha & -\sin\alpha \\ \sin\alpha & \cos\alpha \end{bmatrix}\begin{bmatrix} a \\ b \end{bmatrix}$. This is easy to check by performing the multiplication on the right-hand side of the equation.

Here is a summary of what we have found.

(Algebraic description of rotations around the origin) To get the coordinates of the image of the point (a, b) $\left(\text{or } \begin{bmatrix} a \\ b \end{bmatrix}\right)$ under the rotation around the origin through an angle α, we simply need to multiply the matrix $\begin{bmatrix} \cos\alpha & -\sin\alpha \\ \sin\alpha & \cos\alpha \end{bmatrix}$ with the point $\begin{bmatrix} a \\ b \end{bmatrix}$. □

So far, we have described all of the translations of the plane algebraically, but only very few of the reflections and rotations. For example, we still have not seen an algebraic description of the rotation around the point $(2, 3)$ through $60°$, nor do we know how to encode algebraically the reflection with respect to, say, the horizontal line 1 unit above the horizontal axis. As we will see in the following two examples, that will be surprisingly easy. We will use compositions of symmetries to indicate how the most general cases can be reduced to the few simple cases that we have covered above.

Example 5. Rotations about Any Point

Consider the rotation $\text{rot}(C, 60°)$ about the point $C = \begin{bmatrix} 2 \\ 3 \end{bmatrix}$ through an angle of $60°$. We want to describe that symmetry algebraically. Our solution will be based on the following observation: $\text{rot}(C, 60°) = \text{trans}_v \circ \text{rot}(O, 60°) \circ \text{trans}_{-v}$, where v is the vector starting at the origin and ending at the point C (i.e., it is the position vector of C) and where trans_{-v} denotes the translation along the vector v. This claim is less complicated than it might appear at first glance: we simply say that we can perform the rotation $\text{rot}(C, 60°)$ by first translating the whole plane so that C moves to the origin (that accounts for trans_{-v}, the symmetry that acts first in the composition), then we rotate about the origin through $60°$ (accounting for $\text{rot}(O, 60°)$), and finally, move the whole plane backward along v so that the origin moves back to the point C (accounting for trans_v). A visual explanation is provided in Figures 2.2.7 through 2.2.10.

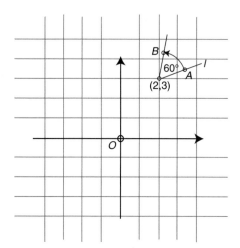

FIGURE 2.2.7 We want to describe the rotation around the point (2, 3) through 60° algebraically. That rotation will move the line *l* to the line passing through the points (2, 3) and *B*, and it will move the point *A* to the point *B*. In the following illustrations, we will keep track of the line *l* and the point *A*, but we should bear in mind that the rotation is applied to all points in the plane.

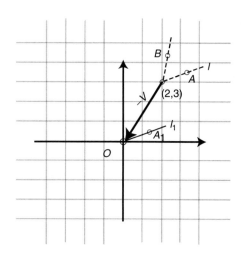

FIGURE 2.2.8 First, we translate the whole plane by the indicated vector. The line *l* is moved to the line l_1 and the points *A* and (2, 3) are moved to the points A_1 and *O*, respectively.

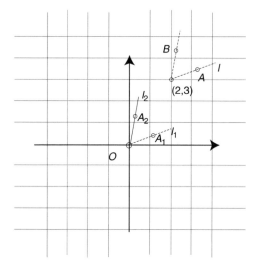

FIGURE 2.2.9 Then we rotate around the point *O* through 60°. The point A_1 is sent to the point A_2 and the line l_1 is moved to the line l_2.

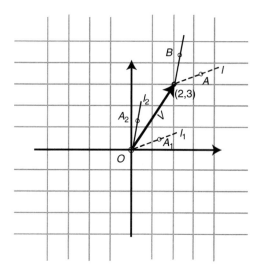

FIGURE 2.2.10 Finally, we translate the plane back through the negative of the vector that we used in the first step. The point A_2 is moved to the point *B* and the line l_2 is moved to the line through (2, 3) and *B*.

The overall effect of the three symmetries we have used in Figures 2.2.8 through 2.2.10 is the same as the rotation of the plane around the point (2, 3). On the surface it appears that we complicated matters, having expressed a simple rotation (around (2, 3) through 60°) in a more complicated way, as a composition of three symmetries: $rot(C, 60°) = trans_v \circ rot(O, 60°) \circ trans_{-v}$. However—and this is precisely our motivation—we have already seen how to encode these three symmetries algebraically, and we can use that to express the starting rotation algebraically. Specifically, the coordinates of the image of any point $\begin{bmatrix} a \\ b \end{bmatrix}$ under $rot(C, 60°) = trans_v \circ rot(O, 60°) \circ trans_{-v}$ are $\begin{bmatrix} \cos 60° & -\sin 60° \\ \sin 60° & \cos 60° \end{bmatrix} \left(\begin{bmatrix} a \\ b \end{bmatrix} + \begin{bmatrix} -2 \\ -3 \end{bmatrix} \right) + \begin{bmatrix} 2 \\ 3 \end{bmatrix}$.

We explain again where this expression comes from: given a point $A = \begin{bmatrix} a \\ b \end{bmatrix}$, $A_1 = \begin{bmatrix} a \\ b \end{bmatrix} + \begin{bmatrix} 2 \\ 3 \end{bmatrix}$ is the image of that point under the translation through the vector $-v$ from the point $\begin{bmatrix} 2 \\ 3 \end{bmatrix}$ to the origin (see Figure 2.2.8 and the associated caption); then $A_2 = \begin{bmatrix} \cos 60° & -\sin 60° \\ \sin 60° & \cos 60° \end{bmatrix} \left(\begin{bmatrix} a \\ b \end{bmatrix} + \begin{bmatrix} -2 \\ -3 \end{bmatrix} \right)$ is the image of A_1 under the rotation around the origin through 60° (see Figure 2.2.9 and the associated caption); and finally, $B = \begin{bmatrix} \cos 60° & -\sin 60° \\ \sin 60° & \cos 60° \end{bmatrix} \left(\begin{bmatrix} a \\ b \end{bmatrix} + \begin{bmatrix} -2 \\ -3 \end{bmatrix} \right) + \begin{bmatrix} 2 \\ 3 \end{bmatrix}$ is the image of A_2 under the translation through the vector v from the origin to the point $\begin{bmatrix} 2 \\ 3 \end{bmatrix}$ (see Figure 2.2.10 and the associated caption).

For example, the image of the point $\begin{bmatrix} 1 \\ 2 \end{bmatrix}$ under $rot(C, 60°) = trans_v \circ rot(O, 60°) \circ trans_{-v}$ is $\begin{bmatrix} \cos 60° & -\sin 60° \\ \sin 60° & \cos 60° \end{bmatrix} \left(\begin{bmatrix} 1 \\ 2 \end{bmatrix} + \begin{bmatrix} -2 \\ -3 \end{bmatrix} \right) + \begin{bmatrix} 2 \\ 3 \end{bmatrix}$. Computing this expression (and recalling the values of $\cos 60°$ and $\sin 60°$) gives:

$$\begin{bmatrix} \cos 60° & -\sin 60° \\ \sin 60° & \cos 60° \end{bmatrix} \left(\begin{bmatrix} 1 \\ 2 \end{bmatrix} + \begin{bmatrix} -2 \\ -3 \end{bmatrix} \right) + \begin{bmatrix} 2 \\ 3 \end{bmatrix} = \begin{bmatrix} \frac{1}{2} & -\frac{\sqrt{3}}{2} \\ \frac{\sqrt{3}}{2} & \frac{1}{2} \end{bmatrix} \left(\begin{bmatrix} -1 \\ -1 \end{bmatrix} \right) + \begin{bmatrix} 2 \\ 3 \end{bmatrix}$$

$$= \begin{bmatrix} \frac{-1 + \sqrt{3}}{2} \\ \frac{-1 - \sqrt{3}}{2} \end{bmatrix} + \begin{bmatrix} 2 \\ 3 \end{bmatrix} = \begin{bmatrix} \frac{-1 + \sqrt{3}}{2} + 2 \\ \frac{-1 - \sqrt{3}}{2} + 3 \end{bmatrix} \overset{(calculator)}{=} \begin{bmatrix} 2.366 \\ 1.634 \end{bmatrix}$$

Example 6. One more Reflection

Question: How can we express, say, the reflection with respect to the horizontal line 1 unit above the horizontal axis algebraically?

Answer: We first translate the line to the x-axis (by means of $trans_v$, where v is the vector $\begin{bmatrix} 0 \\ -1 \end{bmatrix}$), then we reflect with respect to the x-axis, and then we translate back through the vector $-v$. So, to get the image of a point $\begin{bmatrix} a \\ b \end{bmatrix}$, we should first add $\begin{bmatrix} 0 \\ -1 \end{bmatrix}$ to it (accounting for the first translation; see Example 1); then we multiply the resulting point by the matrix $\begin{bmatrix} 1 & 0 \\ 0 & -1 \end{bmatrix}$ (accounting for the reflection with respect to the x-axis; see Example 3); and finally, we add the vector $-\mathbf{v} = \begin{bmatrix} 0 \\ 1 \end{bmatrix}$ to the point obtained in the preceding step (accounting for the last translation; see Example 1). □

Exercise: Fill in the details of the above example.

As we saw in this section, the study of symmetries and their properties can be done through matrices and their algebraic operations. This approach often yields very elegant justifications of various claims that would otherwise need rather clumsy arguments. The theory of matrices has many other applications that are outside the scope of this book. For example, we have hinted in a footnote in Section 1.4 that an explicit formula for Fibonacci numbers could be obtained through the theory of matrices. Another relatively simple application is given in the next subsection.

Matrices and Computers

Each computer graphic is input as a matrix, with entries associated to pixels. For example, the graphic shown in Figure 2.2.11 is in fact a matrix of size 1600×1200, where that product is the number of pixels in the photo. In this black-and-white photo, each positive

FIGURE 2.2.11 The original.

entry in the underlying matrix will generate a black-colored pixel, and each negative entry a white-colored pixel.

Every photo effect is then nothing but a matrix operation. For example, multiplying the matrix with −1 reverses the signs of the entries and the resulting photo effect is the negative of the original (Figure 2.2.12). The effect of one more matrix operation is shown in Figure 2.2.13. (That operation is called convolution of matrices: roughly, in each row of the matrix we replace every entry, starting with the second, with a simple algebraic combination of that entry and the preceding one.)

FIGURE 2.2.12 The negative: multiplying all entries by –1.

FIGURE 2.2.13 Embossed version: a convolution of the matrix.

Exercises:

1. Perform the addition of the following points (given through their coordinates), then plot the given points, as well as the points obtained as a result of the addition.

   a. $\begin{bmatrix} -2 \\ 0 \end{bmatrix} + \begin{bmatrix} 2 \\ 2 \end{bmatrix}$

   b. $\begin{bmatrix} 4 \\ 1 \end{bmatrix} + \begin{bmatrix} -4 \\ -1 \end{bmatrix}$

2. Find the image of the given point under the given translation.

   a. The point is $\begin{bmatrix} 2 \\ 3 \end{bmatrix}$ and the translation is through a horizontal vector of length equal to 2 units.

   b. The point is $\begin{bmatrix} -1 \\ -2 \end{bmatrix}$ and the translation is through the position vector of the point $\begin{bmatrix} -1 \\ -2 \end{bmatrix}$.

3. Find algebraically the image of the point $\begin{bmatrix} 1 \\ 3 \end{bmatrix}$ under the composition of the reflection with respect to the horizontal axis followed by the reflection with respect to the vertical axis.

4. Find algebraically the image of the point $\begin{bmatrix} -1 \\ 1 \end{bmatrix}$ under the composition of the translation through the position vector of the point $\begin{bmatrix} 1 \\ 3 \end{bmatrix}$, followed by the reflection with respect to the vertical axis.

5. Find algebraically the image of the point $\begin{bmatrix} 1 \\ 3 \end{bmatrix}$ under the rotation around the origin through an angle of 45°. (You may want to either recall the values of sin 45° and cos 45° or use a calculator.)

6. a. Find algebraically the image of the point $\begin{bmatrix} 1 \\ 3 \end{bmatrix}$ under the rotation around the point $\begin{bmatrix} 1 \\ 1 \end{bmatrix}$ through an angle of 45°.

   b. Find a formula for the image of any point $\begin{bmatrix} a \\ b \end{bmatrix}$ under the rotation around the point $\begin{bmatrix} 1 \\ 1 \end{bmatrix}$ through an angle of 45°. [*Hint*: see Example 5.]

7. Find a formula for the image of any point $\begin{bmatrix} a \\ b \end{bmatrix}$ under the reflection with respect to the line y = x (consisting of all points with equal coordinates). [*Hint*: express this reflection as a composition of a rotation around the origin through −45° (moving the line y = x to the x-axis), followed by the reflection with respect to the x-axis, and followed by the rotation around the origin through 45°.]

8. a. Consider the transformation dfl_2 of the points in the plane defined as follows: each point A in the plane is moved to the point $dfl_2(A)$ on the horizontal line passing through A, so that the distance between the y-axis and $dfl_2(A)$ is twice than that between the y-axis and A, and so that $dfl_2(A)$ and A are on the same side of the y-axis (see Figure 2.2.14). We note that under the transformation dfl_2, the points on the y-axis do not move. We call this transformation the *deflection* away from the y-axis by a factor of 2. Find a two-by-two matrix such that multiplication of the points in the plane by that matrix is the same as applying the transformation dfl_2 to the points.

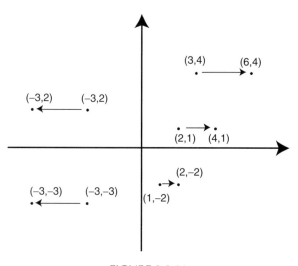

FIGURE 2.2.14

b. The *deflection* away from the vertical axis (the y-axis) by a factor of α, where α ≠ 0, denoted by dfl_α, is defined in a similar way as the deflection from the y-axis by a factor of 2, except that instead of deflecting the points twice farther from the y-axis, we move them α-times farther (or closer, if $\alpha < 1$) from the y-axis. Find a two-by-two matrix such that multiplication of the points in the plane by that matrix is the same as applying dfl_α.

c. Show that deflections from the y-axis preserve lines. That is, show that if A, B, and C are three points on a single line, then $dfl_\alpha(A)$, $dfl_\alpha(B)$, and $dfl_\alpha(C)$ are also on a single line. [*Hint*: assuming B is between A and C, show that $dfl_\alpha(B)$ is between $dfl_\alpha(A)$ and $dfl_\alpha(C)$ by confirming that the distance from $dfl_\alpha(A)$ to $dfl_\alpha(C)$ is the same as the sum of the distances from $dfl_\alpha(A)$ to $dfl_\alpha(B)$ and from $dfl_\alpha(B)$ to $dfl_\alpha(C)$.]

9. Define a transformation as follows: the points on a horizontal line a units above the horizontal axis (the x-axis) are translated a units to the right, whereas the points on a horizontal line a units below the x-axis are translated a units to the left; the points on the x-axis are not moved. We call this transformation the *shear*. Find a matrix such that this shear can be represented by the multiplication of points in the plane by that matrix.

2.3 GROUPS OF SYMMETRIES OF PLANAR OBJECTS

Consider the triple-A design in Figure 2.3.1. If we apply a symmetry, some of the points in the triple-A may end up within the object itself, and some may be sent somewhere outside the object. For example, any translation by a non-zero vector clearly dislodges at least some of the points of the triple-A away from that shape (Figure 2.3.2). But if we apply a rotation around the center of the object through an angle of 120°, then, as is not very hard to see, *all* the points of triple-A fit nicely, and the image of the object is the object itself (Figure 2.3.3).

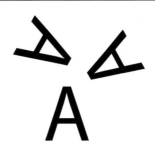

FIGURE 2.3.1 Triple-A design.

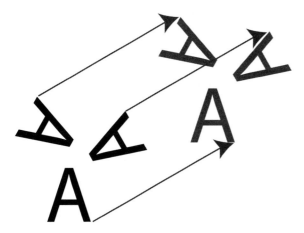

FIGURE 2.3.2 Translations (through nonzero vectors) move some or all the points out of the object.

FIGURE 2.3.3 The rotation around the center of this object and by 120° keeps all the points within the object.

Given an object O in the plane (i.e., given a set of points on the plane), a ***symmetry of the object*** O is a symmetry of the plane that rearranges the points of the object O within O, and such that every position in the object is attained by some point following the rearrangement. The set of all symmetries of an object O is called the **group of symmetries**[*] of the object O. (Pay attention: In general, only some of the symmetries of the plane are symmetries of a given object in the plane.)

Let us find the group of symmetries of the triple-A above. First of all, there is a symmetry that is a symmetry of *every* object in the plane: the identity symmetry. The identity symmetry does not move any of the points in the plane, and, as a consequence, leaves the points in any object confined within that object. Even though this symmetry is not very interesting, we should not forget it when we search for groups of symmetries. We have also noticed one more symmetry of the triple-A: rotation about the center through 120°. With that in mind, it is not hard to see that a rotation about the center by 240° is also a symmetry of the object. There are no other symmetries: we have noticed already that translations will not do, while it is easy to see that no other rotation or reflection is a symmetry of the triple-A. So, we summarize, the group of symmetries of the triple-A is the following set:

{identity, rotation about the center by 120°, rotation about the center by 240°}

We pause to make a note that could help us more easily identify groups of symmetries. Suppose O is an object in the plane and suppose f and g are two symmetries of O. Since f is a symmetry of O, when we apply it to the points of O they are shuffled within O (we do not care here what happens to the other points of the plane). Now, if we take the composition of f followed by the symmetry g, the points of O are still rearranged within O. This is true because we have assumed that g is also a symmetry of O. We conclude the following:

A composition of any two symmetries of any object O in the plane is also a symmetry of that object.

Here is an example that illustrates how this observation could be used. Consider again the triple-A shape from above. We saw initially that rotation f around the center by 120° is a symmetry of that object. Using the above observation, we can now conclude (without looking at the object at all) that the composition $f \circ f$ of f followed by f is also a symmetry. But the composition of a rotation by 120° followed by a rotation by 120° around the same center is a rotation by 240°, and thus we have algebraically discovered the second rotation.

Example 1

We want to find the symmetry group of the object given in Figure 2.3.4. We start with the identity symmetry—as we have noticed, it is a symmetry of every object, and so, in particular, of the one at hand. Next we notice that rotation by 120° about O is also a symmetry (see Figure 2.3.5). Consequently, so is the composition of that rotation with itself, which gives a rotation by 240° about the point O. (Alternatively, we simply observe directly that rotation by 240° is also a symmetry of our object.) Finally, reflection with respect to the lines l_1, l_2, and l_3, as indicated in Figure 2.3.5, also maps the object within itself, and are thus symmetries of our object.

---

[*] The terminology we use here comes from basic modern algebra, where the notion of *group* has a specific meaning; our *groups of symmetries* are groups according to that specific definition.

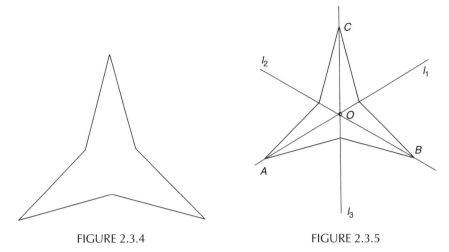

FIGURE 2.3.4   FIGURE 2.3.5

These are the only* symmetries of the object and so the group of symmetries is {identity, rotation about O by 120°, rotation about O by 240°, reflection with respect to l_1, reflection with respect to l_2, reflection with respect to l_3}. Denoting the rotation around a point O and through an angle α by $rot(O, \alpha)$, and denoting the reflection with respect to a line l by $ref(l)$, we can shortly write the group of the symmetry of the object shown in Figure 2.3.4 as follows: {identity, $rot(O, 120°)$, $rot(O, 240°)$, $ref(l_1)$, $ref(l_2)$, $ref(l_3)$}. □

Example 2

Figure 2.3.6 shows a flag. The group of symmetries of the flag consists of the identity symmetry, the reflection with respect to the line l, the reflection with respect to the line m, and the rotation through 180° around the center O (see Figure 2.3.7). Using the shorter notation established in Example 1, we can write that the group of symmetries of the flag is {identity, $rot(O, 180°)$, $ref(l)$, $ref(m)$}. Our symmetries are color-blind (which in this case does not matter much since the colors of the flag are also compatible with the symmetries).

Note that when the two reflections mentioned above are symmetries of an object (and, in particular, of the flag), then it has to be that the rotation about the intersection of the two

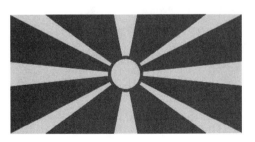

FIGURE 2.3.6

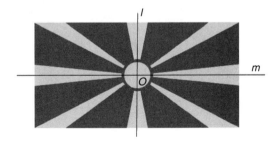

FIGURE 2.3.7

---

* This could be shown more formally by considering some strategically chosen points, and then using the fact that symmetries preserve distances to analyze their images. For example, the points A, B, and C in the picture can only be permuted under any symmetry of the object—or else the distances will not be preserved—and it is easy to see that each of these permutations corresponds to one of the symmetries we have found.

mutually perpendicular lines of reflection O by 180° is a symmetry of that object (whatever the object is). Why? Because the composition of one of the two reflections followed by the other is the rotation about O by 180°—Exercise: Check that claim—and we know that any composition of symmetries of an object is also a symmetry of that object. □

Example 3

See Figure 2.3.8: we assume that the pattern in the row of copies of Dadat *extends unboundedly* on both sides in the same fashion as we show (we can, of course, show only a small part of that unbounded row).

FIGURE 2.3.8

In all of the preceding examples, the groups of symmetries of the objects were finite and none of them contained a translational symmetry. Not so in this case. For example, it is easy to see that there are infinitely many translational symmetries of the pattern. We show the vectors of translation for a few of them in Figure 2.3.9.

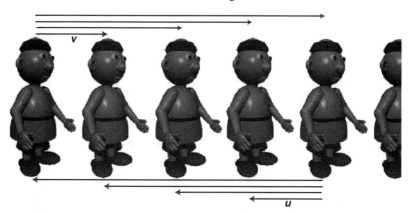

FIGURE 2.3.9

The group of symmetries of the pattern consists of the identity, the translation in the direction of the shortest vector v pointing to the right (shown in Figure 2.3.9), the translation along the vectors twice larger than v (same direction), the translation along the vectors three times larger than v (same direction), and so on; moreover, we have the translation in the direction of the shortest shown vector u pointing to the left, the translation along the vectors twice larger than *u* (same direction), the translation along the vectors three times larger than *u* (same direction), and so on. It is obvious that no non-trivial rotations or reflections are symmetries of this object. □

To simplify our notation, we introduce the notion of *integer **multiples of vectors***. For every integer *n* and every vector *v*, we will denote the vector that is parallel to *v* and that is *n* times longer than *v* by *nv*. If *n* is positive, then we agree that *v* and *nv* have the same direction; otherwise (if *n* is negative), *v* and *nv* have the opposite direction. For example, the vectors shown in Figure 2.3.9 are *v*, 2*v*, 3*v*, and 4*v* (all pointing to the right and in the direction of *v*), and then −*v*, −2*v*, −3*v*, and −4*v* (all pointing to the left and in the direction opposite to that of *v*). The special vector 0*v* is called the ***zero vector*** and it is of length 0: translation along the zero vector is the same as the identity symmetry (no points are moved).

Denoting the translation along the vector *v* by *trans_v*, we can now write the group of symmetries of the pattern indicated in Figure 2.3.8 as follows: {identity, *trans_v*, *trans_{2v}*, *trans_{3v}*, …, *trans_{−v}*, *trans_{−2v}*, *trans_{−3v}*, …}, where the ellipses "…" indicate infinite continuation of the pattern established to their left.

Remember: We are moving points in the *plane*! So, we should resist associating three-dimensionality to the above pictures. For the time being, all we see and do is two-dimensional. Recall also that the arrows we use indicate the direction and length of the translation for *all the points* on the plane. For example, the smallest arrow pointing to the right tells us how *all* the points move, not just the tip of the bald head of one of the copies of Dadat.

We will encounter more patterns like this one in the next section.

Exercises:
1. Find the group of symmetries of each of the letters A, B, E, H, K, M, S, T, V, X, and Z.
2. Find the group of symmetries of each of the three objects shown in Figures 2.3.10 through 2.3.12.

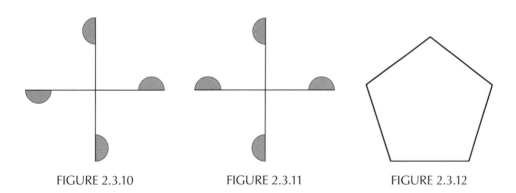

FIGURE 2.3.10   FIGURE 2.3.11   FIGURE 2.3.12

3. Which of the objects shown in Figures 2.3.13 and 2.3.14 is more symmetric. Justify your answer by finding their respective groups of symmetries and comparing their numbers of elements.
4. Find two (types of) bounded objects in the plane with infinite groups of symmetries (an object is *bounded* if it can be enclosed within a square).

FIGURE 2.3.13   FIGURE 2.3.14

5. Find a planar object with (a) exactly three symmetries, (b) exactly four symmetries, and (c) exactly five symmetries.

6. Let l and m be two lines intersecting at a point P at an angle of 30°. Suppose that the reflections with respect to l and m are two symmetries of an invisible object O in the plane. Show that the rotation around the intersection point P and by an angle of 60° is also a symmetry of the object O.

7. Show that there does not exist a planar object such that its group of symmetries consists only of the identity symmetry and two distinct reflections.

8. There are many symmetries of the "sunflower" crop design shown in Figure 2.3.15. Describe them.

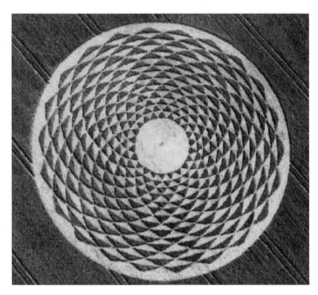

FIGURE 2.3.15 The sunflower crop circle. Photo by Peter Sorensen (2000).

9*. A planar object is *totally disconnected* if no two points can be joined by a curve *within* that object. Find a bounded and totally disconnected object with infinite group of symmetries.

2.4 FRIEZE PATTERNS

Frieze 1 and Some Definitions

We take another look at the object (denoted by O_1 in Figure 2.4.1) considered in the last example in the previous section. Keep in mind that we assume that the row extends unboundedly on both sides and in the same way. Some of the translation symmetries are indicated in Figure 2.4.1.

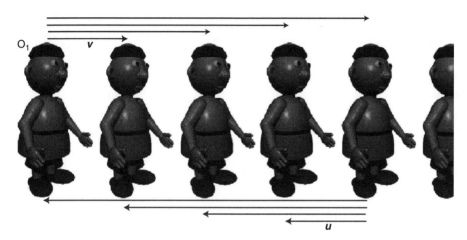

FIGURE 2.4.1 The symmetry group of this pattern, denoted by F_1, has infinitely many translation symmetries.

As we have noticed in Example 3, Section 2.3, all translation symmetries of this pattern have the following property.

(*) The vectors of the translation symmetries are all integer multiples of one fixed vector.

In the pattern shown in Figure 2.4.1, the role of the fixed vector can be taken by any one of the vectors u and v.

Objects in the plane that have translation symmetries satisfying the property (*) are called *frieze patterns*, or *friezes*. Their groups of symmetries are called *frieze groups*.

What kinds of frieze patterns are there? That question is natural, but it has one major weakness: at this point, "*kinds of frieze patterns*" has a somewhat vague meaning. We can be clearer and more specific by considering the groups of symmetries of the friezes, and using them to try to classify the patterns.

Since the structure of the translation symmetries in the frieze groups is fixed by the property (*), it is the rest of the symmetries that change. The part of the group of symmetries consisting of rotations and reflections and their compositions with translations determines the type of the frieze group, and so it determines (in a way) the frieze pattern itself.

For example, consider the group F_1 of symmetries of the object O_1 above. If we omit every second copy of Dadat from O_1, then we get another pattern with *fewer* copies of Dadat. But if we examine the group of symmetries of the latter pattern, we would easily discover that it is very similar to F_1; in particular, there are no other types of symmetries except translations, and these translations will again all be multiples of a single vector. The only difference is that vector would now be twice longer. Since the set of reflection symmetries or rotation symmetries in F_1 is not affected (in both cases there are no reflections and proper rotations), we consider the two frieze patterns (the pattern O_1, and the pattern we get from O_1 by omitting every other copy of Dadat) to be of the same type. ☐

We can now search for *different* frieze patterns according to the discussion above. So far we have the frieze pattern O_1 in the above picture, with the associated frieze group denoted by F_1. Let us take a look at some other examples of frieze patterns and consider their groups of symmetries.

Frieze 2

Consider the pattern shown in Figure 2.4.2 (keeping in mind that all the patterns we consider in this section extend unboundedly on both sides).

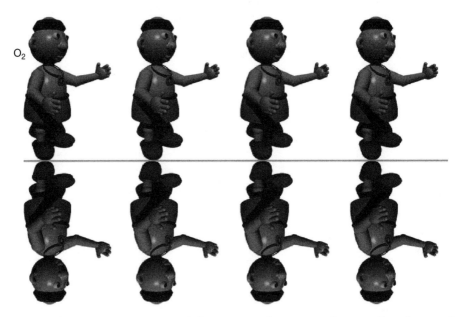

FIGURE 2.4.2 The symmetry group of this pattern has one reflection (beside translations along all integer multiples of one fixed vector). This group is denoted by F_2.

Other than translations we can easily see one more symmetry of O_2: the reflection with respect to the horizontal line we show. Is that all? Not really: we remember that compositions of symmetries of an object must be a symmetry of that object. So, if we compose one of the translations with the reflection, we get other symmetries. More precisely, applying some of our translations and then following that by a reflection with respect to the indicated line is also a symmetry. Recall that compositions of translations and reflections along lines parallel to the vectors of translations are called glide reflections. So, the frieze group F_2 in this case consists of translations and glide reflections. Note that we have not forgotten the identity (and the reflection): it is a translation by the zero vector (followed by reflection, respectively). We see that F_2 is *different* from F_1, so that we have a new frieze pattern according to our criterion. □

Frieze 3

See Figure 2.4.3, ignore the microscopic differences and assume that (the two-dimensional) Dadat together with his clothing, necklace, shadows, and reflections has a bilateral symmetry.

FIGURE 2.4.3 The symmetry group of this pattern is denoted by F_3.

The translations defining frieze patterns according to property (*) are clearly present. There is one more kind of symmetry of this frieze pattern: reflection. There are reflection symmetries of the pattern with respect to the vertical lines passing right through the middle of each copy of Dadat, as well as reflection symmetries with respect to the vertical lines through the midpoint between any two consecutive copies of Dadat. So, in this case, the frieze group F_3 consists of the mandatory translations, as well as the infinitely many reflections we have just described. (Note that since the reflections in this case are with respect to vertical lines, and the latter are not parallel to the vectors of translation symmetries, the composition of a reflection symmetry followed by a translation symmetry is not a glide reflection. Rather, that composition is again a reflection. Can you see why?) □

Frieze 4

The frieze O_4 in Figure 2.4.4 looks like O_2: we have the translation symmetries along integer multiples of a fixed vector (as we should), and we also have glide reflections: for example, translating from A to B, then reflecting with respect to the shown horizontal line produces a glide reflection. (Note that this glide reflection is one of infinitely many glide reflections that are symmetries of O_4.) However, O_4 does not have any reflection symmetries

FIGURE 2.4.4 The symmetry group of this frieze pattern is denoted by F_4.

(unlike O_2, where the reflection with respect to the horizontal line that we depicted was a part of the frieze group of O_2). □

Frieze 5

As before, we ignore the small asymmetries of the front-facing copies of Dadat (Figure 2.4.5).

FIGURE 2.4.5 The symmetry group of this frieze pattern is denoted by F_5.

Besides translations, this frieze pattern has many rotation symmetries. Rotations centered at the points A, B, C, ... (there are infinitely many of these centers of rotation in both directions) and by the angle of 180° are symmetries of this frieze. The group of symmetries F_5 in this case consists of the frieze translations as well as the rotations we have just described. It is obviously different from the previous frieze groups. □

Frieze 6

Ignore the small asymmetries in the images of Dadat.

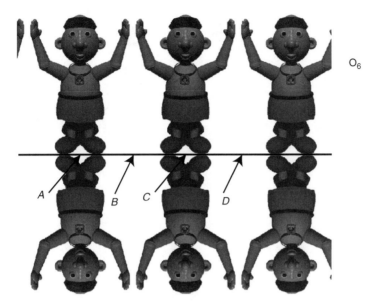

FIGURE 2.4.6 The symmetry group of this frieze pattern is denoted by F_6.

The frieze pattern in Figure 2.4.6 is the most symmetrical one. On top of the defining translations, there are rotations and reflections as symmetries of O_6. Rotations about the points A, B, C, D, ... (extending on both sides) by 180° are all symmetries of the object. So the reflection with respect to the horizontal line is shown. Consequently, compositions of our translations and that reflection (the glide reflections) are also symmetries of O_6. Finally, reflections with respect to the vertical lines through the points A, B, C, D, ... (in both directions) also belong to the frieze group F_6. This is the complete list of symmetries of F_6. So, summarizing, this frieze group F_6 consists of our translations, as well as rotations, reflections, and glide reflections that we have just described. This group of symmetries is obviously not listed above, so we can say that this is indeed a new frieze pattern. □

Frieze 7

As was the case with the preceding frieze, we do have the rotations about A, B, C, ... (infinitely many in both directions) by an angle of 180°. But unlike F_6, the frieze group of symmetries in Figure 2.4.7 does not have the reflection with respect to the horizontal line and it has *fewer* reflections with respect to vertical lines: only the vertical lines through B, D, F, ... (many such in both directions) are lines of reflections belonging to F_7. There is one more type of symmetry of this frieze: the glide reflections obtained by composing one of the translations along the integer multiples of the vector from A to C followed by the reflection with respect to the horizontal line. Note that none of the translations or the

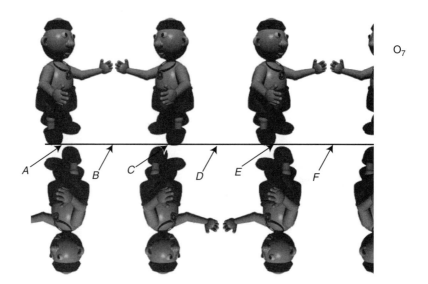

FIGURE 2.4.7 The symmetry group of this frieze pattern is denoted by F_7.

reflection mentioned in the previous sentence are in F_7, but their compositions are there. So, summarizing, F_7 consists of translations, then the rotations about certain points by 180°, reflections along vertical lines, and glide reflections in the horizontal direction. □

At this point our story comes to an abrupt end.

> **Theorem: (The Classification Theorem for Frieze Patterns)** The seven frieze groups F_1, F_2, F_3, F_4, F_5, F_6, and F_7 are the only frieze groups.* Consequently, we can say that the above 7 frieze patterns are the only essentially different frieze patterns.

In the background of the theorem there are precisely defined concepts of frieze groups and frieze patterns, but we have enough explanation and examples to get the main idea.

The theorem tells us that whatever frieze pattern we make, be it very wild or exotic, as long as it fulfils the condition (*) in the definition of frieze patterns, the associated frieze group (of symmetries of that object) must be one of the seven listed above.

In Figure 2.4.8, we show another example of a frieze pattern. It appeared as a crop formation in Beckhampton, England, June 2005 (we assume, as we always have assumed in this section, that the pattern extends unboundedly both to the right and to the left). According to the classification theorem, the symmetry group of this frieze must be the same as exactly one of the seven frieze symmetry groups listed above. Which one? □

---

* *Note 1*: Our notation of frieze groups is not standard, what we have denoted by F_1, F_2, F_3, F_4, F_5, F_6, and F_7 is usually denoted by F_1, F_1^1, F_1^2, F_1^3, F_2, F_2^1, and F_2^2, respectively.

Note 2: The proof of the theorem requires basic group theory (algebra). We note for the readers who are familiar with basic group theory that some pairs of the frieze groups are isomorphic (say, F_1 and F_4).

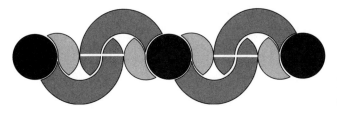

FIGURE 2.4.8 Beckhampton's crop circles.

Start with *any* bounded object and choose a frieze group out of the seven possible. Fix a horizontal line and mark on it the shortest vector of translation, centers of rotations (all rotations are by 180°), and other lines and points related to the symmetries in the frieze group we have chosen (the line with marks could be erased at the end of our construction of a frieze). Then apply repeatedly all of the symmetries in that frieze group to the starting object. The result will be a frieze pattern. We provide the following two simple examples and one that is slightly more complicated. □

Example 1. Using Frieze Groups: A Simple Example

We start with a square (our object) and with the first frieze group F_1. The shortest nonzero vector of translations in F_1 is also depicted in the starting picture.

We assume that the vector we have chosen corresponds to the shortest nonidentity translation in F_1. Now apply all of the symmetries in F_1 to the square shown in Figure 2.4.9. We recall that F_1 consists only of translations along vectors that are integer multiples of the given starting vector. Here is what we get (see Figure 2.4.10).

FIGURE 2.4.9 In this example, we choose the first frieze group F_1.

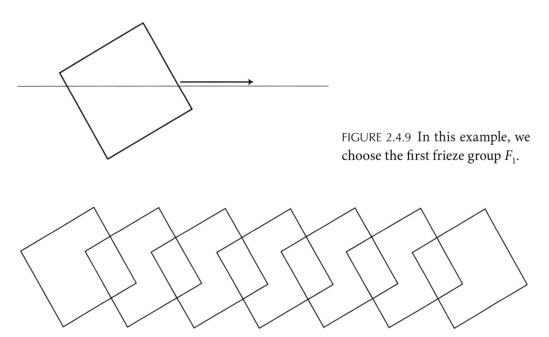

FIGURE 2.4.10 The pattern extends unboundedly in both directions. □

Example 2. Using Frieze Groups: Example 1 Modified

We modify the previous example a bit: this time the starting vector of translation is parallel to one of the diagonals of the square and is exactly one-half of the length of the diagonal (see Figure 2.4.11).

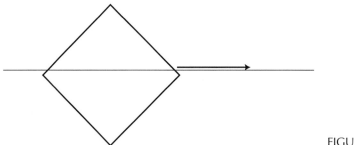

FIGURE 2.4.11

We apply the symmetries in F_1 (i.e., we apply translations along all integer multiples of the vector we depicted), and get the following (see Figure 2.4.12).

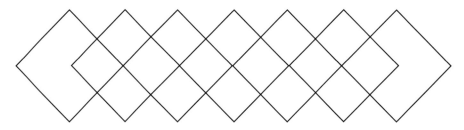

FIGURE 2.4.12 The design obtained by applying F_1 to the square in Figure 2.4.11.

Observe that the frieze we got in this example has an associated frieze group different from F_1 (whereas in the previous example it was the same as F_1). So, when we use frieze groups to design frieze patterns, the patterns we get may have frieze groups of symmetries different from the starting frieze group. □

Exercise: Can you determine which of the seven frieze groups corresponds to the frieze pattern shown in Figure 2.4.12?

Example 3. Using Frieze Groups: A Slightly more Complicated Example

This time we will work with the frieze group F_7. We recall that F_7 contains reflections with respect to vertical lines. Since in this case the smallest nonzero translation vector is twice the distance between any two consecutive vertical lines of reflection, and since the rotations in that group are through 180° and centered in the midpoints between pairs of consecutive vertical reflection lines, all of the symmetries in F_7 are determined by the

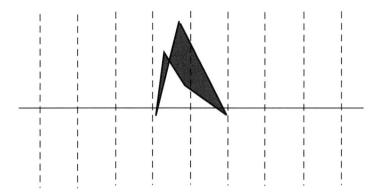

FIGURE 2.4.13 The starting shape and the vertical reflection lines defining the reflections in F_7 are all given. The horizontal line is auxiliary and indicates the direction of the vectors of translation and the centers of rotations. The points of intersection of the shown lines are the centers of rotations in F_7.

positions of the vertical lines of reflections. So, our starting picture (Figure 2.4.13) is an arbitrarily chosen shape and a set of vertical lines (choose them anywhere provided any two consecutive lines are at a fixed distance) indicating the reflections, but also determining the translations and the rotations in F_7.

Now apply all the rotations (centered along the shown horizontal line and in the midpoints between consecutive vertical lines), reflections (with respect to the vertical lines), and translations (the smallest nonzero vector twice the distance between consecutive vertical lines). The result is shown in Figure 2.4.14.

FIGURE 2.4.14 The result of applying the frieze group F_7 to the object shown in Figure 2.4.13.

□

Exercises:
1. Identify the frieze group (one of the seven discussed above) for each of the seven friezes shown in Figures 2.4.15 through 2.4.21.
2. Draw the frieze we get after we apply F_2 to the object O (as shown in Figure 2.4.22).
3. Some frieze patterns cover the entire plane. We show one in Figure 2.4.23. Identify the associated frieze group.
4. In Figure 2.4.24, we show a wallpaper design (we will cover wallpaper designs in the next section). Assuming that the design covers the entire plane, identify within it at least three frieze patterns with different frieze groups.

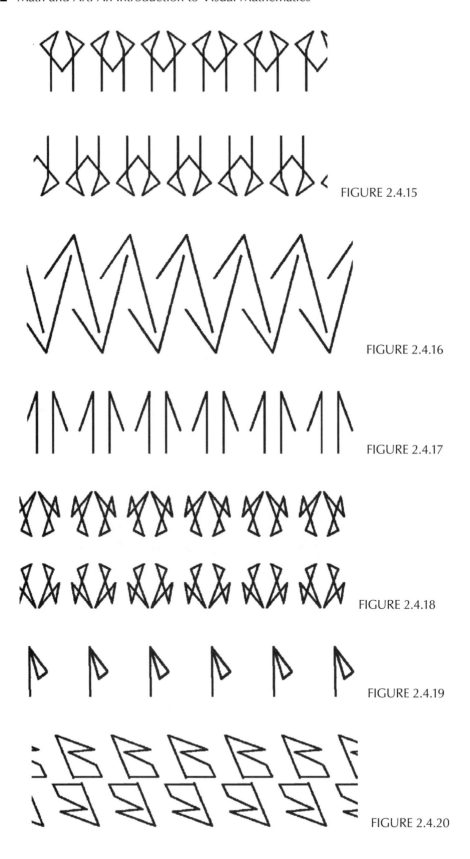

FIGURE 2.4.15

FIGURE 2.4.16

FIGURE 2.4.17

FIGURE 2.4.18

FIGURE 2.4.19

FIGURE 2.4.20

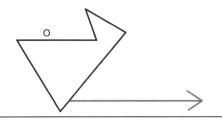

FIGURE 2.4.21

FIGURE 2.4.22 The smallest non-zero vector of translation and the horizontal reflection line are shown.

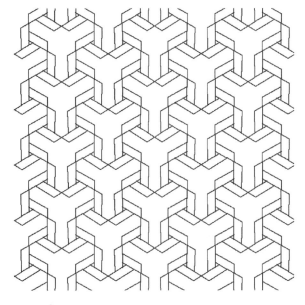

FIGURE 2.4.23 This pattern extends in all directions as indicated here, covering the entire plane.

FIGURE 2.4.24

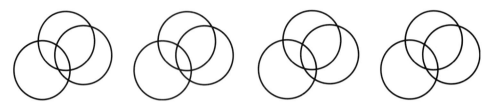

FIGURE 2.4.25

5. The frieze pattern of type F_1 shown in Figure 2.4.25 is made of copies of one circle. Use copies of a single circle to draw the other six types of frieze patterns.

2.5 WALLPAPER DESIGNS AND TILINGS OF THE PLANE

Wallpaper Groups

In the previous section, we looked at (unbounded) objects in the plane (the friezes) with only one type of a translation symmetry: all translation symmetries of the friezes were along vectors that were integer multiples of one fixed vector. As a consequence, most of the frieze patterns extended along rows—so, in some sense, they were one-dimensional. Now we take a look at patterns that have translation symmetries along two nonparallel vectors.

Let us start with a simple example. In Figure 2.5.1, we see a design: it is understood that it extends on all sides so that it extends throughout the entire plane. The two vectors and the letters are not a part of the design: we put them there so that we can explain what is, in this context, the most important feature of this object.

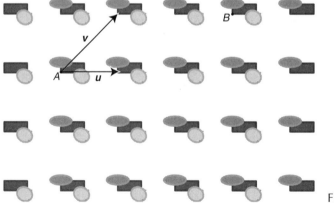

FIGURE 2.5.1

First of all, we notice that translations along any of the two shown vectors are symmetries of this design. Let us call the translations along these two vectors the **basic translations**. Denote the vectors by **u** and **v** as shown in Figure 2.5.1. There are many other translation symmetries in this case. For example, it is not hard to observe that the translation $trans_{AB}$ along a vector extending from A to B is also a symmetry of this object. It is also easy to see that $trans_{AB}$ is equal to the composition of the translation along **u**, followed again by the translation along **u**, and followed by translation along **v** (we remind the reader that two symmetries are equal if the final destinations of the points on the plane are the same—the way the points get there is not important). We can experiment more to see that any translation that is the composition of a translation along an integer multiple of **u**, followed by a translation along an integer multiple of **v** is a symmetry of the design (recall that the negative multiples of a vector point in the opposite direction).

Moreover, any translation symmetry of the object shown in Figure 2.5.1 is a composition of a translation along an integer multiple of **u**, followed by a translation along an integer multiple of **v**. This property is one of the two properties determining the so-called wallpaper designs. An object in the plane is a ***wallpaper design*** if the following two properties are true.

    a. There are two nonparallel vectors such that every translation that can be obtained by composing a translation along an integer multiple of the first vector followed by a translation along an integer multiple of the second vector is a symmetry of the object.

    b. Every translation symmetry of the object must be of the kind specified in (a).

As was the case with frieze patterns, wallpaper designs have also been classified according to their symmetry groups. It turns out that there are exactly 17 types* of symmetry groups of wallpaper designs. So we can say that there are 17 types of wallpaper patterns. Their symmetry groups are somewhat more involved than the symmetry groups of the frieze patterns, so we will be satisfied with just giving one example of a wallpaper design from each of the 17 types. The first design is given in Figure 2.5.1, and the next 16 are depicted in Figure 2.5.2.

The groups of symmetries of wallpaper designs, often called the ***wallpaper groups***, are also known by the name of two-dimensional ***crystallographic groups***. Crystallographers were the first to consider the problem. As the specification "two-dimensional" suggests, there are designs in higher dimensions; in particular, there are three-dimensional patterns with three types of translation symmetries. Curiously, the classification of three-dimensional patterns (or crystals, from the point of view of crystallographers) through their groups of symmetry preceded the above classification of the two-dimensional wallpaper patterns, even though the number of three-dimensional crystallographic groups is much larger: 219 of them. Fedorov was the first to classify both two- and three-dimensional crystallographic groups (the three-dimensional case in 1885; the two-dimensional case in 1891). Mathematicians have covered the four-dimensional case, even though we cannot properly visualize any such pattern. There are 4783 four-dimensional crystallographic groups.

---

* It is beyond the reach of this book to specify precisely what we mean by "types" of symmetry groups. Here is a rough definition for the readers familiar with group theory: two symmetry groups are of the same type if there is a group isomorphism that sends symmetries of a given type (say, rotations) to symmetries of the same type.

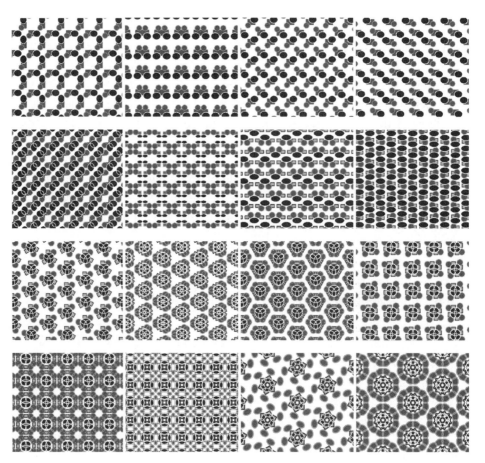

FIGURE 2.5.2 **(See color insert following page 144.)** Wallpaper designs with the remaining 16 types of wallpaper groups.

Tilings of the Plane

Any arrangement of (planar) objects on a plane in such a way that all of the plane is covered and such that any two tiles either share a common vertex (corner), intersect along a pair of their edges, or do not intersect at all, is called a ***tiling*** of the plane. The planar objects we use to cover the plane are called *tiles*. The tiles are usually either polygons or obtained from polygons by bending their edges. The 17 types of symmetry groups can all be realized through tilings of the plane, and so, from the point of view of types of symmetry groups, there are 17 types of tilings of the plane with two types of translational symmetries. We show them all in Figure 2.5.3. Each tiling is denoted by its name according to the crystallographers' notation. We will not go into detailed explanation of the notation beyond noting that the numbers refer to the existence of rotational symmetries (2 for a twofold, or 180°, rotational symmetry, 3 for the threefold or 120° symmetry, and so on), whereas some of the letters refer to other symmetries (*g* stands for glide reflections, *m* for certain other types of reflections).

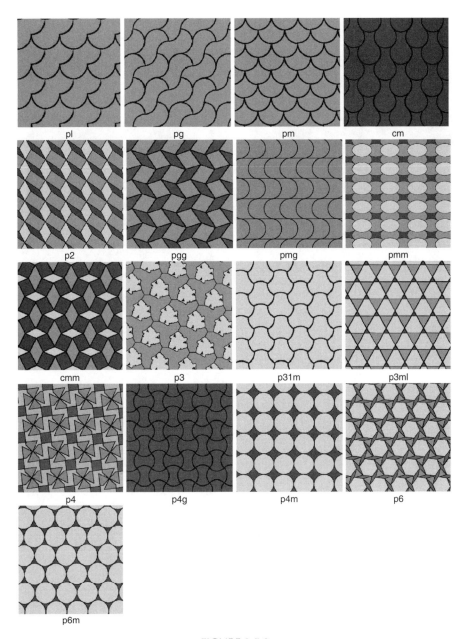

FIGURE 2.5.3

The photos in Figures 2.5.4 through 2.5.6 show three of the above types of tilings.

FIGURE 2.5.4 Roof tiles: tiling pm.

FIGURE 2.5.5 Beehive: tiling p6.

FIGURE 2.5.6 Logs: tiling p6m.

Two artworks based on tilings are shown in Figures 2.5.7 and 2.5.8.

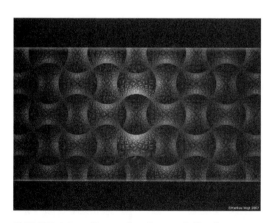

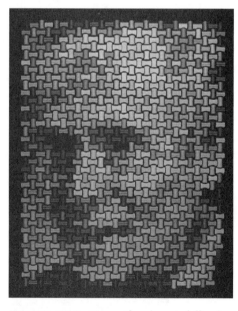

FIGURE 2.5.7 (See color insert following page 144.) Marcus Vogt. *Tiling Shapes 06*, Apophysis fractal generator, then Photoshop, 2007.

FIGURE 2.5.8 (See color insert following page 144.) Ken Knowlton. *Aaron Feuerstein; Spools of Thread*, 32 in. × 26 in. Collection Aaron Feuerstein, © 2001 Ken Knowlton.

Regular Tilings of the Plane

Now we consider regular tilings: a polygonal tiling is **regular** if all the polygons we use as tiles are regular polygons (equilateral triangles, squares, regular pentagons, etc.), if every two adjacent polygons have either a common point or a common edge, and if the polygons around every common point are of the same type throughout the tiling. A regular tiling is **monohedral** if there is just one type of a tile, that is, if the tiling is made of one type of regular polygon of fixed size.

What kinds of monohectral regular tilings are there? In order to answer that question we will do a brief mathematical analysis.

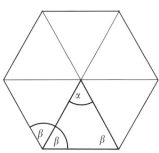

First we compute the interior angle in a regular hexagon and in a regular m-gon, where an m-gon is a polygon with m sides (Figure 2.5.9). We see that α is 360°/6, so (since the sum of the angles in the triangle is 180°) we have that $2\beta = 180° - 360°/6 = 120°$, giving the value of each of the interior angles of the regular hexagon. It is not very hard to see that in case of a regular m-gon the same argument works, except that, rather than dividing by 6 in the first step, we should divide by m. Consequently, the interior angle in a regular m-gon is $180° - 360°/m$.

FIGURE 2.5.9 Computing the interior angles of hexagons.

We digress to point out that in the above argument we have used the α property that the sum of the interior angles in any triangle is 180°. That property is a consequence of the Euclidean postulates, specifically the fifth postulate (see Section 1.1). It is precisely because of that property that there are, as we will see below, very few regular monohedral tiling of the usual plane. Later, in Chapter 5, we will cover another geometry, where the sum of the angles of the triangles are less than 180°, and, as a consequence, where there are infinitely many regular monohedral tilings.

Now, assume that we have a monohedral tiling with n interior angles number of m-gons meeting at each vertex of the tiling. The possibilities are explored in the next paragraph.

A Bit of Math. Classifying Monohedral Tilings

First of all, as we just saw, the interior angle in a regular m-gon is $180° - \frac{360°}{m}$. We have supposed that there are n number of regular m-gons around each vertex in the tiling. So all of the n interior angles starting at that vertex make a complete circle, and thus add up to 360°. So $n\left(180° - \frac{360°}{m}\right) = 360°$. We simplify this equation (first we divide both sides by 360°): $n\left(\frac{1}{2} - \frac{1}{m}\right) = 1$, then we continue in small steps (you should be able to see what has been done in each step; recall that $\Leftrightarrow$ is short for "means the same as"). $\left(\frac{1}{2} - \frac{1}{m}\right) = \frac{1}{n} \Leftrightarrow \frac{1}{2} = \frac{1}{n} + \frac{1}{m} \Leftrightarrow nm = 2m + 2n \Leftrightarrow nm + 4 = 2m + 2n + 4 \Leftrightarrow 4 = 2m + 2n - nm + 4.$

A Bit of Math (contd.)

Now check that $2m + 2n - mn + 4 = (m - 2)(n - 2)$, so that the last equation in the above chain of equations becomes $4 = (m-2)(n-2)$. The only positive integer factors of 4 are 1, 2, and 4; so $m - 2$ and $n - 2$ must be one of these three numbers. We list the cases as follows:

a. $m - 2 = 1$ and $n - 2 = 4$
b. $m - 2 = 2$ and $n - 2 = 2$
c. $m - 2 = 4$ and $n - 2 = 1$

Case (a) gives $m = 3$ and $n = 6$ — that is, we have three ($m = 3$) regular hexagons ($n = 6$) meeting at each vertex.

Case (b) gives $m = 4$ and $n = 4$ — that is, we have four ($m = 4$) regular squares ($n = 4$) meeting at each vertex.

Case (c) gives $m = 6$ and $n = 3$ — that is, we have six ($m = 6$) equilateral triangles ($n = 3$) meeting at each vertex. □

We proved the following theorem.

Theorem: There are exactly three monohedral regular tilings of the plane: one with equilateral triangles (six of them meeting at each vertex), one with squares (four around each vertex) and one with hexagons (three around each vertex).

The three tilings mentioned in the theorem are shown in Figures 2.5.10 through 2.5.12.

FIGURE 2.5.10 Regular triangular tiling.

FIGURE 2.5.11 Regular square tiling (with a camouflaged Dadat).

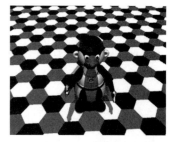

FIGURE 2.5.12 Regular hexagonal tiling (with a camouflaged Dadat).

Archimedean Tilings of the Plane

The number of regular tilings with one or more than one type of tiles is 11. This is an old result: it was first proved by Johannes Kepler in 1619. His paper (in fact, his book) was forgotten for about 300 years, and his results were rediscovered at the beginning of twentieth century. These 11 types of regular tilings are now known by the name of **Archimedean tilings**. Three of them are shown in Figures 2.5.10 through 2.5.12. The remaining eight

Archimedean tilings—these with at least two types of regular polygons used as tiles—are shown in Figures 2.5.13 through 2.5.21. The notation we apply is fairly standard and is almost self-explanatory: for example, 33434 in Figure 2.5.14 means that as we make a circle around a vertex we encounter a triangle, a triangle, a square, a triangle, and a square.

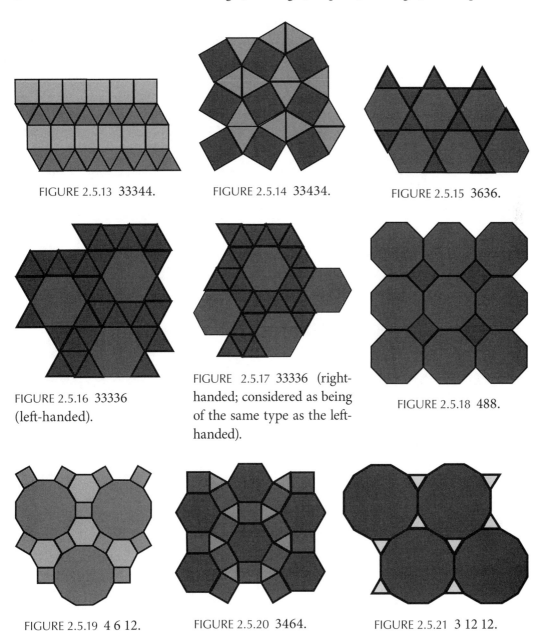

FIGURE 2.5.13 **33344**.

FIGURE 2.5.14 **33434**.

FIGURE 2.5.15 **3636**.

FIGURE 2.5.16 **33336** (left-handed).

FIGURE 2.5.17 **33336** (right-handed; considered as being of the same type as the left-handed).

FIGURE 2.5.18 **488**.

FIGURE 2.5.19 **4 6 12**.

FIGURE 2.5.20 **3464**.

FIGURE 2.5.21 **3 12 12**.

Aperiodic Tilings of the Plane

All of the tilings we have seen so far have translation symmetries. Such tilings are called **periodic tilings**. It is not very hard to construct a (not necessarily regular) tiling that does not have any translation symmetries.

Exercise: Use squares and triangles (any kind of triangles) to construct a tiling of the plane with no translational symmetries. [*Hint*: one row in the tiling should be a pattern made of pairs of adjacent squares followed by triangles, the next row a pattern with triples of adjacent squares followed by triangles, etc.]

Wang (1961) conjectured that if a set of tiles covers the plane, then they could be rearranged so that the new tiling is periodic. The conjecture was disproved by Berger (1966) who used 20426 (sic!) tiles in his construction. He was the first to find an ***aperiodic tiling*** of the plane: a tiling with no translational symmetries and such that no rearrangement of it produces a periodic tiling.[*] We will not show his construction. That number of tiles was decreased several times, the last step so far being done by Penrose (1974), who found a tiling with two tiles that could not be rearranged to become periodic (Figure 2.5.22).[†] We show his tiling below, as well as a tiling by Ammann (also with two tiles (Figure 2.5.23). It is interesting to observe that the tiles in the Penrose tiling (a *dart* and a *kite*) are made by gluing pairs of *golden* triangles of the sort we have dealt with in Section 1.3. (The tiles in the Ammann tiling also have golden proportions!)

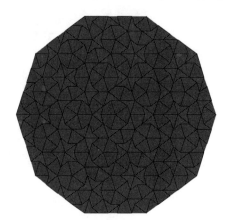

FIGURE 2.5.22 Penrose's aperiodic tiling with kites and darts; each of the tiles is made of two golden triangles.

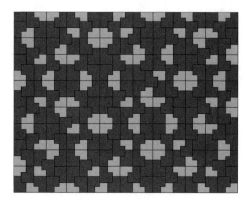

FIGURE 2.5.23 Ammann's aperiodic tiling: the only known aperiodic tiling made of two similar tiles.

Note also that the Ammann tiles are similar, not congruent. It is not known whether there is a tiling with one tile that cannot be rearranged to have translation symmetries.

Exercises:
1. Construct a wallpaper design by repeatedly reflecting the starting object shown in Figure 2.5.24, as well as its reflections, with respect to the horizontal and vertical lines. Find the two basic translations of the pattern.
2. Show directly that there is no regular tiling with regular pentagons as tiles. [*Hint*: the sum of the angles of the tiles around a fixed vertex in the tiling must be 360°.]

---

[*] Sometimes the aperiodic tilings involve certain conditions on how the edges are matched in the tiling.
[†] Both Penrose and Ammann tilings involve certain *matching conditions* regarding the way the tiles must be combined. More information can be found in, say, *Tilings and Patterns*, by Grunbaum and Shephard.

3. Consider the regular tiling 3464 (see Figure 2.5.20). Subdivide each of the regular hexagons into six equilateral triangles. Which tiling from the list of Archimedean tilings do we get?
4. Find four symmetries other than translations for the tiling depicted in Figure 2.5.25.

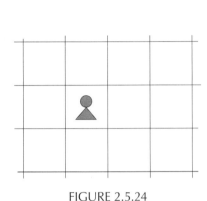

FIGURE 2.5.24

FIGURE 2.5.25 **(See color insert following page 144.)** Floral tiling of the plane, type 4444.

5. a. Tile the plane with the tile shown in Figure 2.5.26.
   b. Tile the plane with the tile shown in Figure 2.5.27.

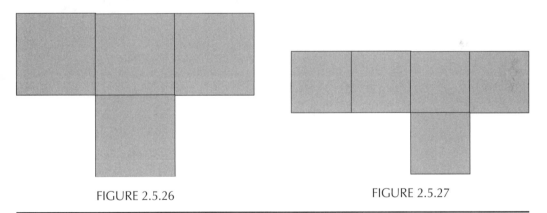

FIGURE 2.5.26

FIGURE 2.5.27

2.6 TILINGS AND ART

The tiling shown in Figure 2.6.1 is a detail of a wall in the Hall of Repose, of the complex Alhambra of Granada. Alhambra was a palace of the Moorish kings of Spain, and its construction started AD 1238. The bird shape of the basic tile is emphasized in Figure 2.6.2, where we modify a detail of Figure 2.6.1.

Many centuries later, in the year of 1936, Maurits Cornelius Escher, a Dutch artist, visited the palace and was impressed and inspired by the mosaics on its walls. Exactly 700 years after the commencement of the construction of the palace, Escher produced the woodcut *Day and Night* (Figure 2.6.4). A detail of the woodcut is shown in Figure 2.6.3, and it clearly points to Alhambra as the source of Escher's inspiration.

FIGURE 2.6.1 A mosaic in the *Hall of Repose* of the palace of Alhambra.

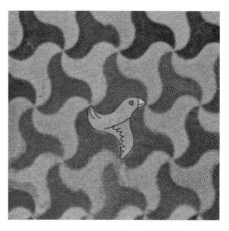

FIGURE 2.6.2 The tiling is monochromatic (one type of a tile) of type p3 (see the classification in the preceding section). We emphasize here the bird-like shape of one of the tiles.

FIGURE 2.6.3 M. C. Escher. *Day and Night*, detail 1938.

FIGURE 2.6.4 M. C. Escher. *Day and Night*, woodcut, 1938.

Two more of Escher's artworks on the theme of the tiling of the plane are shown in Figures 2.6.5 and 2.6.6.

FIGURE 2.6.5 M. C. Escher. *Metamorphosis 1*, woodcut, 1937. This is one of the first prints in the so-called metamorphosis period of Escher (1937–1945). The landscape on the left-hand side of the print is based on Escher's color sketch of the village Amalfi, southern Italy. To the right we see a few types of tilings that continuously morph one into the other.

FIGURE 2.6.6 M. C. Escher. *Reptiles*, lithograph, 1943. The reptile tiling is of type p3 (see Section 2.5); it can be obtained by altering the hexagonal regular tiling. Details are given further in this section.

Tilings, wallpaper designs and frieze patterns have been used as decorations ever since ancient times. One can find them as attributes of the artifacts of every major civilization, decorating walls of the old temples or palaces, or etched in sarcophagi of pharaohs, kings, and warriors. Figures 2.6.7 through 2.6.9 provide a few examples.

FIGURE 2.6.7 Wallpaper design: a detail of ceiling of a tomb, Greece, 1400–1500 BC.

FIGURE 2.6.8 Tiling: a detail of a wine jug, Rhodes, sixth century BC.

FIGURE 2.6.9 Same idea as in Figure 2.6.8: a detail from a marble sarcophagus, Tripoli, Lebanon, second century.

Figure 2.6.10 is a post-Escher piece of art on the same theme.

FIGURE 2.6.10 **(See color insert following page 144.)** John Osborn. *Bats*, ink and watercolor, 1990.

Recipes for Escher-Type Tilings
Recipe 1
Start with a square (we keep an eye on the underlying regular tiling by squares).

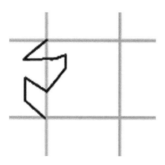 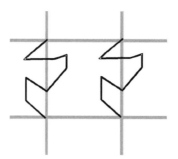 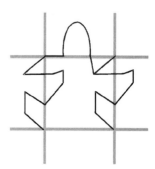

STEP 1. Start with any curve joining the two vertices on the left-hand side of the square.

STEP 2. Translate the broken line as shown.

STEP 3. Now join the two top vertices of the square with any other curve.

Take another look at Step 2: we have connected the two right-hand side vertices of the square in such a way that the curve is a translation of the curve we drew in Step 1, and the part of that curve *within* the central square is symmetric to the part of the curve in Step 1 that is *out* of that square. The same is true for the broken lines in Steps 3 and 4. As long as we have pairs of curves that fulfill these two properties, the tile we make can be used to tile the plane.

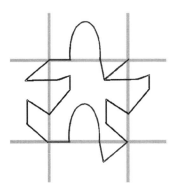

STEP 4. Translate the curve drawn in Step 3 downward as shown.

STEP 5. Draw or color the interior of the contour any way you want.

STEP 6. We do not need the background square grid any more.

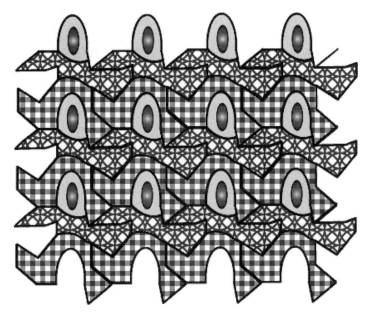

STEP 7. Our construction guarantees that there is a tiling of the plane with our shape as the only tile.

In the last example, the underlying regular tiling was made of one type of tile—a square. The recipe will work for the other two regular monochromatic tilings (triangles and hexagons), as long as the analog of the above-mentioned properties is satisfied. We call these properties *a recipe for Escher-type tilings* and identify them explicitly* as follows:

1. The contour of the tile we construct can be subdivided into pairs of symmetric curves.

---

* The recipe we give is incomplete. The mathematically inclined may want to try to find an example of a tile cooked out of a hexagon and following the recipe, yet not allowing a tiling of the plane.

2. In every pair of symmetric curves identified in 1, the part of the curve *within* the underlying square (equilateral triangle, hexagon) is symmetric to the part of the other curve of the pair that is *out* of the square (triangle, hexagon, respectively).

We note that in the first rule, the pairs of the symmetric curves need not correspond to edges. For example, part 3 of the curve in Figure 2.6.11 is symmetric (via a rotation) to part 4. These two correspond to two edges of the triangle in the background. However, the rest of the curve, corresponding to the top edge of the triangle, is subdivided into two symmetric parts 1 and 2 (again the symmetry is a rotation). The tiling that is generated is shown in Figure 2.6.12.

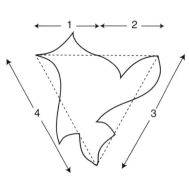

FIGURE 2.6.11

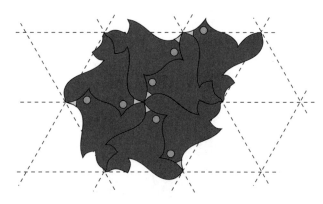

FIGURE 2.6.12 We just need to take copies of that tile, rotate them, and translate them so that they fit.

Recipe 2
We will now modify the first example to identify the freedom we have under our recipes for Escher-type tilings.

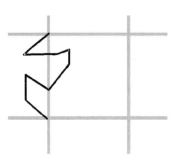

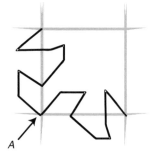

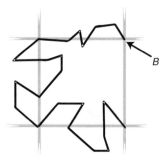

STEP 1. We start with any curve joining the left-hand side vertices.

STEP 2. This time we rotate the curve around the point A and by –90° (negative sign since we go clockwise). Notice that the two curves satisfy both properties in the recipe for Escher-type tilings.

STEP 3. We draw any curve joining the two top vertices of the square. In the next step we rotate this curve around B through 90°.

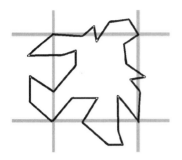  

STEP 4. We get another curve and the last two curves satisfy the two properties in the recipe.

STEP 5. We erase the background, ...

STEP 6. ... and we draw or color whatever we want inside our tile.

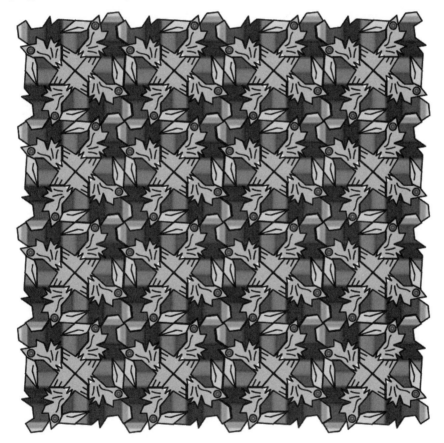

STEP 7. Rotate and translate our tile until the copies we get fit nicely, and cover the plane—our construction guarantees that such a cover of the plane exists.

Exercises:
1. Start with the monohedral regular tiling made of equilateral triangles. For each triangle in the tiling, identify the center point and connect them with the vertices of the triangle. Then erase the starting regular tiling. What do we get? Is the resulting tiling regular?
2. In Figure 2.6.13, we show the first two steps of tiling Escher's lithograph *Reptiles* (Figure 2.6.6).

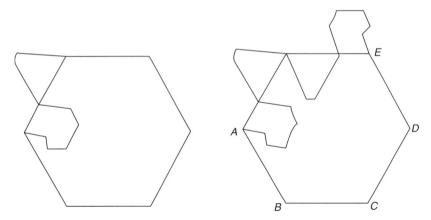

FIGURE 2.6.13 The first step in the tiling of Escher's *Reptiles* (Figure 2.6.6).

Complete the tile by modifying the other two pairs of edges, then tile the plane as in Escher's graphics.

3. Start with an equilateral triangle and find the midpoint of each of the three sides. Draw a curve from the midpoint of one of the sides to one of the two vertices on that side, then rotate a copy of it around the midpoint by 180°. Repeat this procedure for the other three sides of the triangle. Then tile the plane with the object you get.

4. In this exercise we will deal with Voronoi diagrams. A *Voronoi diagram* of a set *S* of points in the plane is a tiling of the plane such that the points in each tile are those that are closest to one of the points *S*. For example, in Figure 2.6.14 we show some points on two lines in the plane, and in Figure 2.6.15 we show the associated Voronoi diagram.

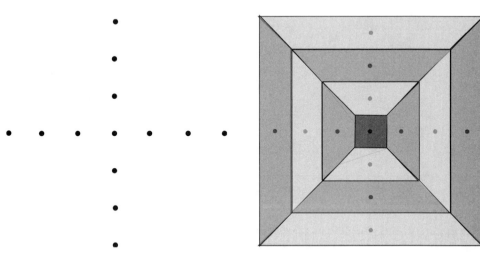

FIGURE 2.6.14 A set *S* of points in the plane: the pattern extends without end on all four sides.

FIGURE 2.6.15 The tiling from the Voronoi diagram for the set *S*. Each tile is the set of points closest to one of the points in *S*.

Find the Voronoi diagram for the set of vertices of each of the three regular monohedral tilings of the plane.

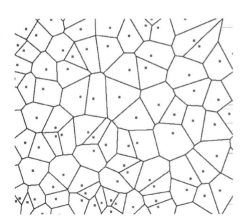

FIGURE 2.6.16 The Voronoi diagram of a set of random points in the plane.

FIGURE 2.6.17 **(See color insert following page 144.)** James J. Lemon: *voroscreen006*, 2006.

Voronoi diagrams have esthetic potential already noticed by some artists. Figures 2.6.16 and 2.6.17 show one of them.

CHAPTER 3

Similarities, Fractals, and Cellular Automata

3.1   SIMILARITIES AND SOME OTHER PLANAR TRANSFORMATIONS

Recall that symmetries are exactly those plane transformations that preserve distances. Of course, not all plane transformations preserve distances. Here is one example.

Example 1. A Similarity

Fix a point O on the plane and define a plane transformation g by declaring that for every point A on the plane, the image $g(A)$ is the point on the half-line from O toward A that is at twice the distance from O to A (see Figure 3.1.1). Note that O is the only point that does not move under g.

Obviously, g does not preserve distances, for the distance between O and A is twice smaller than the distances between their images $g(O)$ and $g(A)$. So, at least for this particular pair of points (O and A) the distance increased by a factor of 2. What about other pairs of points in this example? Let us see: we take another point B and compare the distances AB and $g(A)g(B)$, where, by our definition of g, the point $g(B)$ is on the semiline from O to B but twice farther from O than the point B (see Figure 3.1.2). What is the ratio $\dfrac{g(A)g(B)}{AB}$ of the distances between the indicated points? The triangles OAB and $Og(A)g(B)$ are similar since they have the same angle at the vertex O, and since the ratios of the sides $\dfrac{Og(A)}{OA}$ and $\dfrac{Og(B)}{OB}$ are equal (being, by assumption, both equal to 2). Consequently, the other two sides, $g(A)g(B)$ and AB, are also proportional and with the same coefficient of proportions. That means $\dfrac{g(A)g(B)}{AB}$ is also equal to 2.   □

We have just shown that the transformation g in the above example has the following property: for every two points A and B on the plane, $g(A)g(B) = 2AB$. We call such transformations *similarities*.

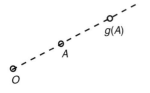

FIGURE 3.1.1

A plane transformation *f* is a ***similarity*** if there exists a *positive* number α such that for every two points *A* and *B* on the plane, we have $f(A)f(B) = \alpha AB$. In words, the distance between the points $f(A)$ and $f(B)$ is α times the distance between the points *A* and *B*. The number α is called the ***stretching factor*** of the similarity *f*.

As was the case with symmetries, similarities are also both onto and one-to-one (hence, they are bijections; Exercise 6).

Two remarks are in order. First, we call the number α the stretching factor even when it is between 0 and 1, in which case the associated similarity *shrinks* distances. For example, if $\alpha = 1/2$, then the distance between any two points is decreased by a factor of 2 after we apply the similarity (see Example 2). Second, we note that when $\alpha = 1$, then the similarity preserves distances, and so it is a symmetry. Consequently, symmetries can now be considered as special similarities: they are similarities with the stretching factor equal to 1.

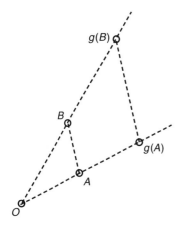

FIGURE 3.1.2

Example 2. A Central Similarity

Again, fix a point *O* and define a similarity *f* by declaring that for every point *A* on the plane, the image $f(A)$ is the point on the half-line from *O* toward *A* that is at twice closer to *O* than point *A*. We could repeat the argument in Example 1 to check again that for *every* two points *A* and *B*, the distance $f(A)f(B)$ is twice smaller than the distance *AB*. So, this transformation is also a similarity, and its stretching factor α is 1/2 (see Figure 3.1.3; keep in mind that we only show what *f* does to two points, *A* and *B*; all of the points on the plane except the center *O* are moved in this way). ☐

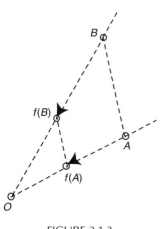

FIGURE 3.1.3

Examples 1 and 2 are of the same type: we fix a point *O* and then we change the distances from the other points to *O* by a fixed factor without moving them away from the rays originating from *O*. Such similarities are called ***central similarities*** (or ***dilations***); the point *O* is called the ***center of the central similarity***. We already know that there are similarities that are not central similarities—for example, translations or reflections. We also know that reflections and translations are symmetries. There are similarities that are neither symmetries nor central similarities, and we will now give an example of such similarity.

Example 3. A Spiral Similarity

In a nutshell, we take a central similarity f centered at a point O and with a stretching factor of, say, 1.5, and we compose it with a rotation g around O by 60°. Denote the resulting transformation by h. A more detailed explanation is given in illustrated Steps 1 and 2 as follows:

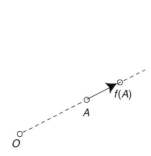

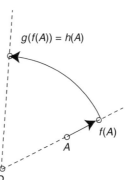

STEP 1. First we apply the central similarity f. Every point A is moved to $f(A)$, 1.5 times farther from O.

STEP 2. Then we rotate the point $f(A)$ by 60° around O. The resulting point is the image of the point A under the composition h of f followed by g.

Since the central similarity f increases distances by a factor of 1.5 and since rotations do not change distances at all, it follows that the composition mapping h must change distances by a factor of 1.5 too, so that h is a similarity with a stretching factor of 1.5. It is not a central similarity since the points and their images are not on the same semiline originating at O, the only fixed point of h. □

Let us play a bit with h before we proceed with the main line. We take any point A different from O and then we repeatedly apply the similarity h. We plot $h(A)$, then we plot the image $h(h(A))$ of $h(A)$ under h, then $h(h(h(A)))$, and so on. What do we get? In Figure 3.1.4, we show A and the 12 consecutive images that we get after applying the similarity h up to 12 times to the point A.

We recognize the shape of a spiral. The points you see are, spiralwise, A, then $h(A)$, then $h(h(A))$, ..., all the way to $h(h(h(h(h(h(h(h(h(h(h(h(A))))))))))))$! The last one is the rightmost point. We will come back to the subject of repeated procedures in Sections 3.3 through 3.5.

This is one of the reasons why similarities of the same type as h (central similarities followed by rotations—details in a couple of paragraphs) are called spiral similarities.

FIGURE 3.1.4

We pursue this subject one step further. Figure 3.1.5 again depicts a point and the points we get from it by repeatedly applying a modification of the mapping h, which we will call h^*. This time we take the stretching factor in h^* to be 1.01 (instead of 1.5) and the angle of the rotation h^* to be $360°/\phi$ (recall that ϕ stands for the golden ratio). The reason we take 1.01 is practical—more points will fit in a reasonably sized square box. We take $360°/\phi$ because of aesthetic reasons—all is clear from Figures 3.1.5 and 3.1.6.

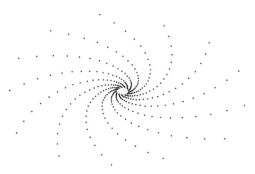

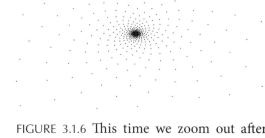

FIGURE 3.1.5 Start with a point and then apply h^* exactly 310 times. We see many spirals. How many? Do you recall the Fibonacci numbers?

FIGURE 3.1.6 This time we zoom out after applying h^* 910 times. Now we see two types of spirals dancing the Fibonacci dance: exactly 13 spirals winding clockwise and 21 counterclockwise.

Except for the bounding circle shown in Figure 1.4.12 and the sizes of the points, Figures 1.4.12 and 3.1.6 depict the same two-dimensional object.

The concept of consecutive applications of spiral similarities is illustrated in the artwork in Figure 3.1.7.

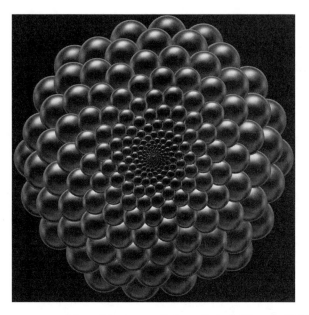

FIGURE 3.1.7 Marc Thomson, *Sphere Spiral Cluster*, 2003.

We now back up and continue our main storyline: we want to know what kind of similarities are there. We have described two spiral similarities in the last examples. Formally, a **spiral similarity** is a composition of a central similarity centered at a point O (with some stretching factor) followed by a rotation around the same point O (and by some angle). When this rotation is by an angle of $0°$ (i.e., when it is the identity symmetry), then the spiral similarity reduces to a central similarity. It turns out that aside from symmetries, spiral similarities are one of only two kinds of similarities. The similarities of the other type are called dilative reflections. A **dilative reflection** is a composition of a central similarity followed by a reflection with respect to a line passing through the center of the central similarity. An example is as follows:

Example 4. A Dilative Reflection

We compose a central similarity f with center O and with a stretching coefficient of 3, followed by a reflection g with respect to the line l passing through O (see illustrated Steps 1 and 2 below). The resulting transformation is a dilative reflection.

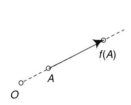

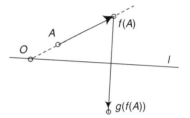

STEP 1. First apply the central similarity f. Every point A is moved to $f(A)$, three times farther from O.

STEP 2. Then reflect with respect to the line l. The dilative reflection h sends the point A to the point $g(f(A))$, and operates similarly on every other point in the plane. □

Here is the promised classification of all similarities of the plane:

> **Theorem:** Every similarity is a symmetry, a spiral similarity, or a dilative reflection.

Since we already know what kinds of symmetries are there, this theorem gives us a complete description of all similarities.

A Proof of the Classification Theorem for Similarities

One way to justify this theorem is to produce an argument similar to the one we gave for the classification theorem for symmetries in Section 2.1. The main difference in this case is that in place of the translation that we used in the justification given in Section 2.1, we use a central similarity. A sketch of the main idea in one case is given in Figure 3.1.8.

Initially we are given a similarity f, the vertices A, B, and C, and the images $f(A)$, $f(B)$, and $f(C)$ of these vertices under f. In the case depicted in Figure 3.1.8, we show how to find a spiral similarity that sends A, B, and C to $f(A)$, $f(B)$, and $f(C)$, and that is equal to the starting similarity f.

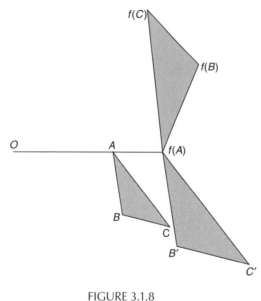

FIGURE 3.1.8

First identify O on the line joining A and $f(A)$ by using the proportion $\dfrac{f(A)f(B)}{AB} = \dfrac{Of(A)}{OA}$.

Then apply the central similarity of stretching factor $\dfrac{Of(A)}{OA}$; it moves $\triangle ABC$ to $\triangle f(A)B'C'$. In the final step, rotate around the point $f(A)$ so that $\triangle f(A)B'C'$ is moved to $\triangle f(A)f(B)f(C)$.

If the orientations of the vertices of the triangles $f(A)B'C'$ and $f(A)f(B)f(C)$ (in the order they appear in this notation) were opposite (one clockwise, the other counterclockwise), then instead of rotating around $f(A)$ as in the last step above, a reflection with respect to a line passing through $f(A)$ would be needed to send $\triangle f(A)B'C'$ to $\triangle f(A)f(B)f(C)$. $\qquad\square$

Objects in the plane that can be mapped one onto the other by means of a similarity are called similar objects. We can say that similar objects have the same shape but (in general) different sizes. The converse is also true: if two plane objects have the same shape but possibly different sizes, then there is a similarity that maps the points on one of the objects onto the points of the other object. In Figures 3.1.9 and 3.1.10 we show two similar objects.

FIGURE 3.1.9 Same shape, different sizes.

FIGURE 3.1.10 "Same shape" includes the possibility of opposite orientation.

Some Other (Partial) Transformations of the Plane

A transformation is ***partial*** if it does not act on some of the points of the plane. This means that for every partial transformation f there is a point A in the plane such that $f(A)$ is not defined at all. We will now describe a few new transformations and some partial transformations. We will encounter or use them later.

a. *The Squaring Transformation*

Fix a point O on the plane, a semiline x starting at the point O (see Step 1 below), and a point X on the semiline x (we need it to simplify the notation only). Define a transformation s as follows: $s(O) = O$, and for any other point A, we first double the angle XOA (denoted by α in the illustrated Step 2) and then we choose the point $s(A)$ on the new semiline such that the distances OA and $Os(A)$ satisfy the equation $Os(A) = (OA)^2$ (see Step 2 below). We call s the **squaring transformation**; the reasons for choosing this terminology are given in the next (optional) section.

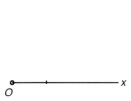

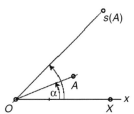

STEP 1. We start with a point O, a semiline x starting at O and a segment of unit length.

STEP 2. For every point A other than O, we double the angle α, and we mark a point $s(A)$ on the new ray of the doubled angle such that the distance $Os(A)$ is the square of the distance OA.

Note that s is not a similarity (so, it is not a symmetry either). Similarities preserve angles (Exercise 4), whereas the transformation s does not preserve angles: the angle α is changed after applying s to the points on the two semilines making that angle. (However, it turns out that α is more of an exception than a rule: only the angles at O can be changed after applying s—the other angles do not change their magnitude.*) ☐

b. *The Squaring Transformation Composed with a Translation*

We compose the squaring transformation s with a translation (the composition is denoted by m). So, the initial setting is the same as for the transformation s, except that this time we also have a fixed vector determining the translation. The details are shown in the illustrated Step 1 below.

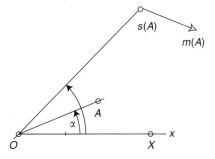

STEP 1. First apply s to any point A as described in the previous example. This will give us the point $s(A)$. Then we translate $s(A)$ by a fixed vector given in advance. The transformation that moves each A to the point $m(A)$ as described in this diagram defines our transformation m.

We will encounter the transformation m later in this chapter. ☐

---

* This claim enters the realm of complex functions—*analytic complex functions* preserve angles except at the points where they are not *conformal* (complex analysis, late years undergraduate course).

c. Circular Inversion

We start with a circle, and we map the points inside the circle to the points outside the circle, and, conversely, the points outside the circle to points inside the circle. A precise description is given below (illustrated Steps 1 and 2). The mapping that we describe below is called a **circular inversion**, shortly **inversion**,* and is denoted by *inv*. It is also called a reflection in a circle.

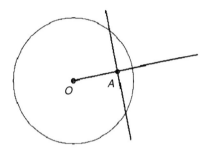 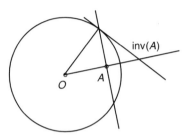

STEP 1. We choose any point *A* inside the given circle, we connect it to the center *O* of the circle and then construct a perpendicular to the line *OA*.

STEP 2. Identify one of the two intersecting points between the perpendicular constructed in the previous step and the circle. Construct the tangent line to the given circle passing through that intersection point. The intersection of the tangent and the line *OA* is the image of the point *A* under this transformation. It is denoted by *inv*(*A*) in this figure.

To get the image of a point *B* outside the circle, follow the above procedure in reverse: first find the tangent line from *B* to the given circle, then construct the perpendicular from the tangential point to the line *OB*; where this perpendicular intersects *OB* is the image point *inv*(*B*). (All this means that if you put *B* in place of *inv*(*A*) in the above illustration, then *inv*(*B*) will be where *A* is.)

Notice that the points on the circle are not moved by the inversion *inv*. Also notice that the center point *O* has *no* image at all. Consequently circular inversions are partial transformations.

We will deal with inversions more extensively later on (Section 4.2). □

d. Linear Fractional Inversion

A composition of a circular inversion (*inv*), followed by a reflection with respect to a line passing through the center of the circle will be called here the **linear fractional inversion** (see the illustrated Step 1 below, where we denote the linear fractional inversion partial transformation by *m*). So, to get *m*(*A*) we first apply the inversion with respect to the

---

* The terminology in classical geometry and in complex analysis is conflicting, and in the latter case, the word "inversion" is used to denote the *linear fractional inversion* that we will encounter in the next example.

given circle to get inv(A) (as explained above), then we reflect with respect to a given line l passing through the center O of the circle. As was the case with inversions, the point O is not sent anywhere in the plane, and so this transformation is also partial.

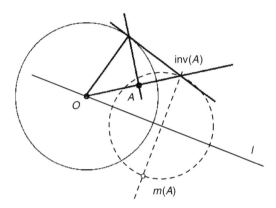

STEP 1. The linear fractional inversion: the point A is moved to the point $m(A)$.

Exercises:
  1. Determine whether the figures given below are always, sometimes, or never similar.
     a. A large circle and a small circle
     b. A large ellipse and a small ellipse
     c. Two rectangles
     d. Two golden rectangles
     e. Two pentagons
     f. Two regular pentagons
     g. A regular pentagon and a regular hexagon
     h. Two isosceles triangles.
  2. Determine which of the following statements are true and which are false.
     a. Any two isosceles triangles must be similar.
     b. Any two golden triangles must be similar.
     c. Any two acute golden triangles must be similar
     d. Every pentagon is similar to itself.
     e. If A and B are similar and B and C are also similar, then A and C must be similar.
  3. The object shown in Figure 3.1.11 is an example of a *fractal* called the Sierpinski triangle (the general notion of a *fractal* will be defined in a couple of sections). As we see, the Sierpinski triangle is obtained by removing the triangle obtained by connecting the midpoints of the sides of the largest triangle, and then repeating that procedure ad *infinitum* (infinitely many times) to the smaller triangles we get in each step. Describe one central similarity with a stretching factor not equal to 1, which moves the points of the Sierpinski triangle within itself. (In order to describe a central similarity you would need to identify its center and its stretching factor.)

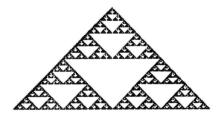

FIGURE 3.1.11 Sierpinski triangle.

4. Let A be a rectangle of size a by b such that when A is folded into two rectangles, the resulting rectangles are similar to the original. What must be the ratio of $a:b$? (This is the underlying principle behind the European "A-size" paper.)

5. Show that similarities preserve angles. This means that for every three points A, B, and C on the plane, and for every similarity f, the angle ABC is congruent to the angle $f(A)f(B)f(C)$.

6. Show that every similarity is both one-to-one and onto.

7. Suppose f is a central similarity centered at a point O and suppose g is a rotation centered at O. Show that the composition of f followed by g is the same as the composition of g followed by f.

8*. We are given two points A and B and we also know where their images $f(A)$ and $f(B)$ under a spiral similarity f are. Find the center of the spiral similarity.

9. Find the images of the points $(1, 1)$, $(0, 1)$, and $(-1, 1)$ (given in Descartes coordinate system; see Section 2.2) under the squaring transformation s. Conclude that points on a single line are not necessarily mapped to points on (another) single line. This also shows that the squaring transformation is not a similarity.

Extra: Yet Another Transformation

FIGURE 3.1.12 Hans Holbein, *The Ambassadors*, 1533. There is a mysterious object by the bottom of the painting.

FIGURE 3.1.13 The mysterious object takes the feature of a skull after we *shear* the points in the plane, as shown here.

3.2* COMPLEX NUMBERS (OPTIONAL)

Preliminaries

As we saw in Section 2.3, rotations and reflections can be encoded as matrix multiplication. It is not very hard to see that central similarities can also be expressed in terms of matrix multiplication. For example, the central similarity f, centered at the origin and with a stretching factor α, can be described using matrices as follows. Given any point $\begin{bmatrix} a \\ b \end{bmatrix}$ in the coordinate plane, the image of that point under the central similarity f is $\begin{bmatrix} \alpha & 0 \\ 0 & \alpha \end{bmatrix} \begin{bmatrix} a \\ b \end{bmatrix} = \begin{bmatrix} \alpha a \\ \alpha b \end{bmatrix}$, and so the action of the central similarity f is the same as multiplication (of the points in the plane defined through their coordinates) by the matrix $\begin{bmatrix} \alpha & 0 \\ 0 & \alpha \end{bmatrix}$. Every other central similarity can be represented as a composition of a translation (bringing the center of the central similarity to the origin), followed by a central similarity with the center at the origin, followed by a translation (bringing the origin back to the original center of the starting similarity). The starting central similarity can thus be represented in terms of the matrix operations we have used to encode each of the three similarities in that composition.

The following is then a natural question to ask: what kinds of onto and one-to-one transformations (i.e., bijective transformations) can be described as multiplication by matrices? It can be shown that the reflection with respect to the line passing through the origin and making an angle of 45° with the x-axis, the deflections (encountered in Exercise 8, Section 2.2), the shear (encountered in Exercise 9, Section 2.2), and the compositions of these three types of transformations are the only such transformations that can be expressed as multiplications by matrices.* These transformations are *linear*, meaning that they move points on a line to points on a line. So, nonlinear bijective transformations cannot be expressed as multiplication by matrices. Thus, for example, the squaring transformation, which we have described in Section 3.1, cannot be described in terms of a multiplication by a matrix, since it is not linear (Exercise 8, Section 3.1). Consequently, we need a new theory if we want to encode this transformation algebraically. The basic theory of complex numbers is one way to tackle this problem, and we present it in the rest of this section. We point out that the theory of complex numbers has many other applications that are beyond the scope of this book.

Basics of the Theory of Complex Numbers

We will now describe the most basic elements of the theory of complex numbers. The theoretical machine that we will build, as elementary as it is, will be sufficient to elegantly encode the squaring transformation and many other planar transformations, including similarities. The construction of the *field* of complex numbers that we will present here is not difficult, but the reader should be prepared for a few abstract notions.

The first and the most important notion that starts the theory of complex numbers is that of the imaginary unit denoted by $\sqrt{-1}$. It is a new "*number*" (or an abstract entity) with the following defining property: its square is equal to -1. We will denote it by i. So, $i = \sqrt{-1}$ stands for a new entity with the property that $i^2 = -1$.[†]

A complex number is a formal algebraic combination of the type $a + bi$, where, as we have postulated, i stands for the imaginary unit $\sqrt{-1}$, and where a and b are (usual) numbers. The addition sign "+" is just a symbol, but there is no error if we view this "+" as an extension of the usual addition of (usual) numbers. Similarly, we should view the

---

* This is a consequence of a basic theorem in linear algebra (that every invertible matrix is a product of elementary matrices) and of the observation that multiplications of points (through their coordinates) by elementary matrices determine reflections, deflections (with respect to the y-axis or x-axis), or shears.

† In the natural scheme of constructing complex numbers, there is one more complex number with the same property: $-i = -\sqrt{-1}$.

operation bi as an extension of usual multiplication. The number a in the complex number $z = a + bi$ is called the **real part** of z, and is usually denoted by Re(z), while b is the **imaginary part** of z, denoted by Im(z).

The so-called *field of complex numbers* is then determined through the following definition of **addition** and **multiplication** of complex numbers:

> **Addition:** $(a + bi) + (c + di) = (a + c) + (b + d)i$
>
> **Multiplication:** $(a + bi)(c + di) = (ac - bd) + (bc + ad)i$

While the addition formula is *natural* and simple, the multiplication seems a bit mysterious. We can clear away some of that mystery by noting that the formula is obtained just by expanding the product on the left-hand side the usual way and by replacing i^2 by -1. Specifically: $(a + bi)(c + di) = ac + bci + adi + bdi^2 = (ac-bd) + (bc + ad)i$. So, there is no need to memorize the multiplication formula: to find the result of the multiplication of two complex numbers, we just need to multiply $a + bi$ by $c + di$ in a natural way, and use $i^2 = -1$ whenever we encounter i^2. A link between multiplication of complex numbers and multiplication of 2×2 matrices and points (written in columns) will be given at the end of this section.

Here are a couple of examples to illustrate these two definitions:

$$(2 - 3i) + (-4 + 2i) = (2 - 4) + (-3 + 2)i = -2 - i$$

$$(2 - 3i)(-4 + 2i) = (2)(-4) + (2)(2i) + (-3i)(-4) + (-3i)(2i) = -8 + 4i + 12i - (6)(-1)$$
$$= -2 + 16i$$

A reader with a sharp eye might have noticed that we have used $2-3i$, an expression we have not formally defined. For practical reasons we will have to avoid being too formal, hoping that some minor conventions will be tacitly accepted. In the mentioned example, it is understood that $2 - 3i$ is short for the more formal $2 + (-3)i$.

The operations of subtraction and division of complex numbers are offshoots of the above definitions of addition and multiplication of complex numbers. We show below how these are done. The computations between the beginning and the end of each of the two formulas given below are for the sole reason of explaining the origins of the formulas in terms of the basic operations of addition and multiplication of complex numbers. This is especially important to keep in mind in case of the rather *ugly* division formula. We should remember the very first step in the longish computation (multiplication by $\frac{c - di}{c - di}$), and the rest of it follows naturally.

> **Subtraction:** $(a + bi) - (c + di) = (a + bi) + [(-c) + (-d)i] = (a - c) + (b - d)i$
>
> **Division:** $\dfrac{a + bi}{c + di} = \dfrac{a + bi}{c + di} \dfrac{c - di}{c - di} = \dfrac{(a + bi)(c - di)}{c^2 - (di)^2} = \dfrac{(ab + bd) + (bc - ad)i}{c^2 + d^2}$.
>
> $$= \dfrac{(ab + bd)}{c^2 + d^2} + \dfrac{(bc - ad)}{c^2 + d^2}i$$

(The division is defined only if at least one of c and d is not equal to 0.)

We illustrate this as follows (note the first step in the division example):

$$(2 - 3i) - (-4 + 2i) = [2 - (-4)] + (-3 - 2)i = 6 - 5i$$

$$\frac{2 - 3i}{-4 + 2i} = \frac{2 - 3i}{-4 + 2i}\frac{-4 - 2i}{-4 - 2i} = \frac{1}{(-4)^2 + 2^2}(2 - 3i)(-4 - 2i) = \frac{1}{20}(-14 + 8i)$$

$$= -\frac{14}{20} + \frac{8}{20}i$$

(Do not forget that $i^2 = -1$. We have used it a couple of times above.)

Since every complex number $a + bi$ is uniquely determined by the pair (a, b), visualizing complex numbers can be done in the same manner as visualizing pairs of numbers. We illustrate this in Figure 3.2.1 (compare with Figure 2.2.1).

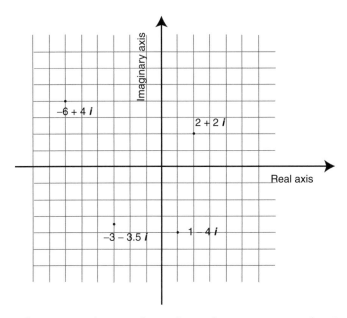

FIGURE 3.2.1 Visualizing complex numbers: the real part measures the signed distance to the vertical axis, and the imaginary part is the signed distance to the horizontal axis.

Now, we can start to express plane transformations in terms of operations of complex numbers. This really means that, given a transformation t of the points of the plane, we will find an expression $f(z)$, such that the image $t(A)$ of any point A in the plane can be obtained by finding the complex number z associated to A (as described above), applying the expression $f(z)$ to that complex number, and finally identifying the point corresponding to the complex number $f(z)$. The process will be further clarified in the examples given below.

Example 1. Translations

At this point, we can easily describe any translation in terms of algebraic operations of complex numbers. Since there is no substantial difference between adding $a + bi$ and $c + di$, and adding $\begin{bmatrix} a \\ b \end{bmatrix}$ and $\begin{bmatrix} c \\ d \end{bmatrix}$, it follows from what we have seen in Section 2.2 that a translation by a vector v is the same as addition by a fixed complex number that corresponds to that point in the plane for which *v* is the position vector. So, if we are given that the vector *v* is the position vector of the point (a, b), then $f(z) = z + (a + bi)$ defines that translation in terms of complex numbers. For example, $f(z) = z + (2 + 2i)$ (addition by the complex number $2 + 2i$ shown in Figure 3.2.1) is the same as translation through the vector starting at the origin and ending at the point corresponding to $2 + 2i$. We summarize: $f(z) = z + (a + bi)$ is the translation by the position vector of the point (a, b). □

Example 2. A Special Reflection

It is also easy to describe the reflection with respect to the horizontal axis in terms of complex numbers since such reflections (as we have seen in Section 2.2) merely change the sign of the second coordinate of each point in the plane. Consequently, the corresponding change of the associated complex numbers is given by the algebraic transformation that sends each complex number $z = a + bi$ to the complex number $a - bi$. The complex number $a - bi$ is called the **conjugate** of the complex number $z = a + bi$ and it is denoted by $\bar{z}$. So, we have $f(z) = \bar{z}$, where $\bar{z}$ is the reflection with respect to the horizontal axis. □

For the rotations and other more complicated plane transformations, we need to slightly expand our theory.

The distance between a (point corresponding to a) complex number $z = a + bi$ and the origin of the coordinate system is called the **modulus** of the complex number, and it is denoted by $\|z\|$. It follows from the Pythagorean theorem that $|z| = \sqrt{a^2 + b^2}$ (see Figure 3.2.2). Since translations preserve distances, the distance between two complex numbers z and w is the same as the distance between their images $z - z$ and $w - z$ under the translation by $-z$. Since $z - z$ is obviously 0, it follows from the above interpretation of modulus that the distance between 0 and $w - z$ is $|w - z|$. Summarizing this part, the distance between the complex numbers z and w is the same as $|w - z|$.

The angle between the ray starting at origin O and ending at z, and the positive (rightward) direction of the horizontal axis is called the **argument** of z. It is denoted by θ in Figure 3.2.2.

We will now use the modulus and the argument of z to write this complex number in another form. First we complicate a bit and write $z = a + bi = |z|\left(\dfrac{a}{|z|} + \dfrac{b}{|z|} i\right)$, then we notice by looking at the triangle in Figure 3.2.2 that $\dfrac{a}{|z|} = \cos\theta$ and $\dfrac{b}{|z|} = \sin\theta$. So, we can write $z = |z|(\cos\theta + \sin\theta\, i)$.

This way of representing z allows us to see more easily what happens when we multiply all complex numbers by a fixed complex number. We fix a complex number z with an argument θ and modulus $|z|$. As we saw above, we can now write $z = |z|(\cos\theta + \sin\theta i)$. Take any other complex number w, say, with modulus $|w|$ and argument α, so that $w = |w|(\cos\alpha + \sin\alpha\, i)$.

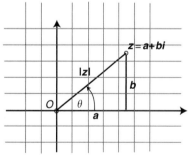

FIGURE 3.2.2 The modulus $|z|$ and the argument θ of $z = a + bi$.

We now multiply z and w (keeping in mind that all the numbers we see are just ordinary numbers, with the exception of the imaginary unit, for which we have $i^2 = -1$):

$$[|z|(\cos\theta + \sin\theta\, i)][|w|(\cos\alpha + \sin\alpha\, i)]$$
$$= |z||w|[\cos\theta\cos\alpha - \sin\theta\sin\alpha + (\cos\theta\sin\alpha + \cos\alpha\sin\theta)i]$$

We now recall the following two trigonometric identities: $\cos\theta\cos\alpha - \sin\theta\sin\alpha = \cos(\theta + \alpha)$ and $\cos\theta\sin\alpha + \cos\alpha\sin\theta = \sin(\theta + \alpha)$. So, we can write: $|z||w|[\cos\theta\cos\alpha - \sin\theta\sin\alpha + (\cos\theta\sin\alpha + \cos\alpha\sin\theta)i] = |z||w|[\cos(\theta + \alpha) + \sin(\theta + \alpha)i]$.

In summary, $[|z|(\cos\theta + \sin\theta\,i)][|w|(\cos\alpha + \sin\alpha\,i)] = |z||w|[\cos(\theta + \alpha) + \sin(\theta + \alpha)i]$. In words, this means that multiplication of complex numbers by a fixed complex number z amounts to multiplying their moduli by $|z|$, and adding their arguments to the argument θ of z. An illustration is provided in Figure 3.2.3.

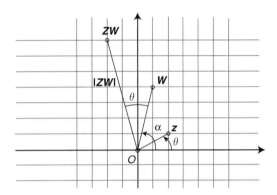

FIGURE 3.2.3 To get the point corresponding to the product of a complex number w by a fixed complex number z, we need to add their arguments (i.e., we need to rotate w by the argument of z, denoted by θ in this figure), and then multiply the modulus of w by the modulus of z (i.e., we need to change the distance to O by a factor of $|z|$).

We are now ready to describe all the transformations we have encountered so far.

Example 3. Rotations

We start with rotations around the origin. It follows from our description of multiplication by a fixed complex number (see, for example, the caption of Figure 3.2.3) that multiplication of all complex numbers by a fixed complex number of modulus 1 and argument α is the same as rotating around the origin about the angle α. Consequently, we have the following: $f(z) = (\cos\alpha + i\sin\alpha)z$ is the rotation around the origin through the angle α.

For example, rotation by $60°$ around the origin is the same as multiplication of all complex numbers by the complex number $\cos 60° + i\sin 60° = \frac{\sqrt{3}}{2} + \frac{1}{2}i$, and can thus be expressed by $f(z) = \left(\frac{\sqrt{3}}{2} + \frac{1}{2}i\right)z$. We can now easily find the image of any point under this rotation. For example, the image of the point $(2,3)$ is $f(2 + 3i) = \left(\frac{\sqrt{3}}{2} + \frac{1}{2}i\right)(2 + 3i) = \sqrt{3} + i + \frac{3\sqrt{3}}{2}i - \frac{3}{2} = \sqrt{3} - \frac{3}{2} + \left(1 + \frac{3\sqrt{3}}{2}\right)i$, that is, the point $\left(\sqrt{3} - \frac{3}{2}, 1 + \frac{3\sqrt{3}}{2}\right)$. □

Since rotations with other centers of rotation can be described as compositions of a translation (bringing the center to the origin), followed by a rotation around the origin, followed by a translation (bringing the origin back to the original center of the starting rotation), we can describe them in terms of operations of complex numbers by using the descriptions of translations and rotations around origin done in this section. We will not do this in detail; we just point out that the idea is the same as the one used in Example 4 in Section 2.2.

Example 4. The Composition of a Translation and a Rotation

We pause for a moment to illustrate how we can use complex numbers to show very easily that $trans_v \circ rot(O, \alpha) = rot(V, \alpha) \circ trans_v$, where $V = trans_v(O)$ (V is where the origin O goes after applying the translation). As we know, in order to show that these two compositions are equal, we need to show that every point in the plane is moved to the same point by both compositions. First we recall that $rot(O, \alpha)$ is the same as multiplication by the complex number $\lambda = (\cos\alpha + i\sin\alpha)$, and that translation through v is the same as addition by the complex number corresponding to the point V (since v is the position vector for the point V). Moreover, because of the argument given in the preceding paragraph, $rot(V, \alpha)$ is the same as addition by the complex number $-v$ (translating the point V to the origin), followed by the multiplication by $\lambda = (\cos\alpha + i\sin\alpha)$ (rotating around the origin), followed by addition with the complex number v (translating the origin back to the point V). So, if A is any point, and if z_A denotes the associated complex number, according to what we have just said, we have that $rot(V, \alpha)(A) = \lambda(z_A - v) + v$.

Now we can compute $trans_v \circ rot(O, \alpha)(A) = trans_v(\lambda z_A) = \lambda z_A + v$. We do similar computation for the action of $rot(V, \alpha) \circ trans_v$:

$$rot(V, \alpha) \circ trans_v(A) = rot(V, \alpha)(z_A + v) = \lambda(z_A + v - v) + v = \lambda z_A + v.$$

So, $trans_v \circ rot(O, \alpha)$ and $rot(V, \alpha) \circ trans_v$ indeed move the point A (or z_A) to the same position. □

Example 5. Central Similarities Centered at the Origin

These are very easy to encode. The central similarity centered at the origin and with a stretching factor β acts in the same way as multiplication of all complex numbers by the number β. This means $f(z) = \beta z$ is the central similarity centered at the origin and with a stretching factor β. □

Exercise: Use Examples 3 and 5 to express the multiplication of all complex numbers by a fixed complex number as a composition of a rotation around the origin followed by a central similarity centered at the origin.

Example 6. The Squaring Transformation

The squaring transformation s is described in Section 3.1. Given our visual description of multiplication of complex numbers (Figure 3.2.3), we now see that the squaring transformation is simply the multiplication of each complex number by itself. So, we have that $s(z) = z^2$ is the squaring transformation.

The reason for calling s the "squaring transformation" is now clear. □

Example 7. The Squaring Transformation Followed by a Translation

This transformation (denoted by m in Section 3.1; we will denote it by j here, saving m for the transformation in Example 8 below) is a composition of a squaring transformation followed by a translation. The former, as we just saw, is described by $s(z) = z^2$, while we found earlier that the latter is addition by some fixed complex number $a + bi$. So, j is defined by $j(z) = z^2 + (a + bi)$ for some constants a and b chosen in advance. Summarizing, $j(z) = z^2 + (a + bi)$ is the composition of the squaring transformation followed by a translation through the position vector of the point (a, b).   ☐

We will need the transformation j to describe some nice-looking *fractals* in a couple of sections.

Example 8. The Linear Fractional Inversion

See Example (d) in Section 3.1 for the definition of the linear fractional inversion. We will assume here that the circle in that example is of radius 1, the center of that circle is at the origin, and the line of reflection is the x-axis. The setting is shown in Figure 3.2.4.

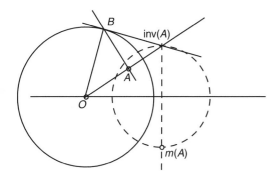

FIGURE 3.2.4 The same illustration was seen in Section 3.1, except that here we assume that the circle centered at O is of radius 1, and the horizontal line is along the x-axis. This figure shows how the transformation moves the points inside the unit circle; for other points refer to the last part of Section 3.1.

The triangle with vertices at O, A, and B is similar to the triangle with vertices at O, B, and $inv(A)$: they are both right-angled triangles and they have a common angle at O. Consequently, their sides are proportional. So, $\dfrac{OA}{OB} = \dfrac{OB}{O\,inv(A)}$ and recall that the circle is of radius 1 (so, $OB = 1$), then we have $OA = \dfrac{1}{O\,inv(A)}$. Now, if we consider A and $inv(A)$ as complex numbers, the last equation tells us that the moduli of these two complex numbers are mutually reciprocal. Also notice that the arguments of (the complex numbers corresponding to) A and $m(A)$ (i.e., their angles with the x-axis) are mutual negatives.

We claim that this complicated transformation can be expressed by $m(z) = \dfrac{1}{z}$, where z stands for the complex number corresponding to the points in the plane.

Here is a brief justification. If $z = |z|(\cos\theta + i\sin\theta)$, then $\dfrac{1}{z} = \dfrac{1}{|z|}\dfrac{1}{\cos\theta + i\sin\theta} = \dfrac{1}{|z|}\dfrac{1}{\cos\theta + i\sin\theta}\dfrac{\cos\theta - i\sin\theta}{\cos\theta - i\sin\theta} = \dfrac{1}{|z|}\dfrac{\cos\theta - i\sin\theta}{\cos^2\theta + \sin^2\theta} = \dfrac{1}{|z|}[\cos(-\theta) + i\sin(-\theta)]$, where we have used $\cos^2\theta + \sin^2\theta = 1$, $\cos(-\theta) = \cos\theta$, and $\sin(-\theta) = -\sin\theta$. Summarizing, $\dfrac{1}{z} = \dfrac{1}{|z|}[\cos(-\theta) + \sin(-\theta)i]$. This means that the modulus of $m(z) = \dfrac{1}{z}$ is the reciprocal of the modulus of z and that its argument is the negative of the argument of z, and so it indeed corresponds to the image $m(A)$ of the point A corresponding to the complex number z, as indicated in Figure 3.2.4. □

Example 9. Inversion

Comparing the circular inversion described in Example c, Section 3.1 and the linear fractional inversion, it is clear that we can obtain the inversion with respect to a circle by composing the linear fractional inversion with respect to the same circle with the reflection with respect to the horizontal axis. Since reflections with respect to horizontal axis are conjugations of complex numbers (see Example 2, this section), it follows that $inv(z) = \frac{1}{z}$ is a description of the inversion in terms of complex numbers. □

Matrices and Multiplication of Complex Numbers

Recall from Example 1 that adding complex numbers to a fixed complex number is basically the same as adding points to a fixed point. Since points (written as columns) are 2×1 matrices, we see that addition by a fixed complex number is the same as addition by a fixed 2×1 matrix. Multiplication of complex numbers by a fixed complex number can also be described in terms of matrices.

Fix a complex number $a + bi$. According to the multiplication formula, we have $(a + bi)$ $(x + yi) = (ax - by) + (bx + ay)i$. Now, let us multiply the matrix $\begin{bmatrix} a & -b \\ b & a \end{bmatrix}$ and the point $\begin{bmatrix} x \\ y \end{bmatrix}$: $\begin{bmatrix} a & -b \\ b & a \end{bmatrix} \begin{bmatrix} x \\ y \end{bmatrix} = \begin{bmatrix} ax - by \\ bx + ay \end{bmatrix}$. Comparing the result of this multiplication with one of the multiplication of the complex numbers, the following is obvious: the points corresponding to these two results are the same, and in both cases, we get the point with the first coordinate $ax - by$ and with the second coordinate $bx + ay$. We may, thus, conclude that the effect of multiplication of all complex numbers by the fixed complex number $a + bi$ is the same as the effect of multiplying all points (written in columns) by the fixed matrix $\begin{bmatrix} a & -b \\ b & a \end{bmatrix}$.

This link could have been used to deduce some of the theory in this section from what we have covered in Section 2.2. For example, we know from Section 2.2 that rotation about the origin and through an angle α is the same as the multiplication of the points in the plane by the matrix $\begin{bmatrix} \cos\alpha & -\sin\alpha \\ \sin\alpha & \cos\alpha \end{bmatrix}$. The above short analysis now implies that the multiplication by the matrix $\begin{bmatrix} \cos\alpha & -\sin\alpha \\ \sin\alpha & \cos\alpha \end{bmatrix}$ has in turn the same effect as multiplication by the complex number $\cos\alpha + i\sin\alpha$ (simply take $\cos\alpha$ in place of a and $\sin\alpha$ in place of b in the matrix $\begin{bmatrix} a & -b \\ b & a \end{bmatrix}$). So, we have arrived at the conclusion of Example 3 in this section that multiplication by $\cos\alpha + i\sin\alpha$ has the effect of rotating the plane around the origin through the angle α.

This analysis also relates the multiplication of matrices mentioned in the preliminaries of this section with the multiplication of complex numbers by a fixed (real) number covered in Example 5, both defining central similarities of the plane. □

Exercises:

1. Perform the following operations, ending with a complex number in the form $a + bi$.
   a. $(2 + 3i)(1 - i) + 5 + 6i$
   b. $\dfrac{1 - i}{2 - i} - 2$

2. Find the modulus and the argument of the complex number $\dfrac{\sqrt{3}}{2} + \dfrac{i}{2}$.

3. Describe in terms of operations of complex numbers the transformation of the plane that is the composition of the reflection with respect to the x-axis, followed by a translation by the position vector of the point (1,1). [*Hint*: your answer should be of the form $f(z) =$ _____, where the blank space should be filled with some expression involving z.]

4. Consider the expression $f(z) = (\cos 30° + i \sin 30°)z + 2$. What symmetry does f define?

5. Consider the expression $f(z) = 3(z + 2)$. Describe the similarity determined by f as a composition of a central similarity and a translation.

6. Find the expression $f(z)$ that describes the rotation about the point (1,1) through the angle of 30° in terms of complex numbers. [*Hint*: express this rotation as a composition of a translation, followed by a rotation around the origin, followed by another translation.]

7. Find the image of the point $A = (3, 3)$ under the following transformations.
   a. The squaring transformation
   b. The linear fractional transformation

Figure 3.2.5 depicts a set of complex numbers described in its caption, and Figure 3.2.6 is an artwork on the same theme.

FIGURE 3.2.5 Some of the points corresponding to the complex numbers $2n \cos\left(\dfrac{2\pi m}{n} + n\right) + i2n \sin\left(\dfrac{2\pi m}{n} + n\right)$, where n ranges from 1 to about 150, and, for each such n, m ranges from 1 to $n - 1$.

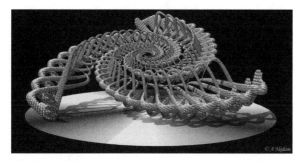

FIGURE 3.2.6 Arend Nijdam. ~*XD Spiral Sculpture*, Xenodream 1.5, 2006.

3.3 FRACTALS: DEFINITION AND SOME EXAMPLES

A *fractal* is an object O possessing the property of *proper* self-similarity. This means that a part A_1 of O is similar to a proper part A_2 of A_1. In case of planar objects, we can describe fractals as follows: O is a fractal if there is a (planar) similarity that sends a part A_1 of O onto a proper part A_2 of A_1. Note that this similarity cannot be a symmetry since we require that the image A_2 of A_1 be a proper part of A_1.

The optimal case is when A_1 is the same as the object O, in which case we say that O is a *complete fractal*. The majority of fractals we will encounter will not be complete.

Let us reflect on this definition for a moment. Suppose A_1 is similar to its proper part A_2 via a similarity f. Then f moves the points of A_1 to the points of A_2. Since A_2 is a part of A_1, all of A_2 (viewed as a part of A_1) is moved via f onto a part of A_2. Denote that part of A_2 by A_3. Then, since the similarity f moves A_2 onto A_3, these two are similar. So we have A_1 similar to A_2, which in turn is similar to A_3. We can continue this process further to get a part A_4 of A_3 similar to A_3, then a part A_5 of A_4 similar to A_4, and so on. We get an infinite sequence of objects $A_1, A_2, A_3, A_4, A_5, \ldots$, all of them similar and such that A_1 contains A_2, which in turn contains A_3, which in turn contains A_4, and so on.

We note before we continue that our definition of a fractal is rather weak since it includes a wide variety of not very interesting objects. For example, each line segment is a fractal since it is certainly self-similar to a proper part of itself. It follows that every object that contains a line segment is also a fractal. We will focus our attention to what we will subjectively consider to be *interesting fractals*.

Example. A Fractal

In Figure 3.3.1, we see a simple fractal. We will use it to illustrate the above analysis. The "head" A_1 of this object is made of infinitely many embedded squares, and each successive smaller square is twice that of shorter edges. The second largest square, together with all of the smaller squares within it, is denoted by A_2. It is a twice smaller copy of A_1 and it is similar to A_1 under the central similarity f centered at the joint central point of all of the squares and with a stretching factor of 1/2. This central similarity f sends A_2 to the part A_3 that is twice smaller than A_2. This process continues without end.

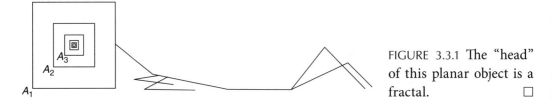

FIGURE 3.3.1 The "head" of this planar object is a fractal. □

A fractal, one may loosely say, is an object where *the greater elements are arranged like the parts, and the parts like the greater elements*. It is a curious fact that the statement in italics is taken verbatim from an English translation of a manuscript printed in the first half of the eighteenth century. The author, Emanuel Swedenborg, was a prominent scientist (and mystic) of his time and in the italicized statement he referred to heaven.

Trees

Our initial goal is to produce a few basic interesting fractals. We are fully aware that the concept "interesting" is almost entirely subjective, but since we do want to sieve out trivial fractals (as is a simple line segment, say), we will try to trace the boundary between *interesting* and *trivial* through a number of examples. We start immediately by constructing a simple (interesting) tree fractal.

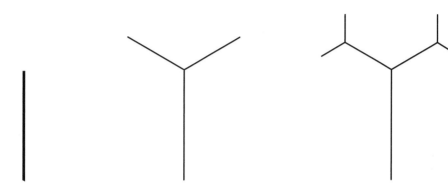

STEP 1. The initial object is a simple vertical line segment.

STEP 2. Branch the upper part of the line into two parts, 60° on both sides of the trunk. The ratio of the lengths of the branches and the trunk is 1:2. This step defines the rest of the procedure and, ultimately, the fractal itself.

STEP 3. Perform exactly the same procedure to the two new branches to get four more branches. The outcome is a tree-like object. It is NOT a fractal yet, but it starts exhibiting some self-similarity: the right-hand branch together with the smaller branches *looks like* the whole tree.

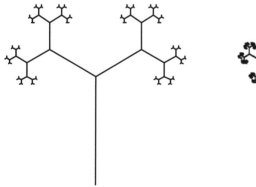

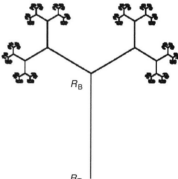

STEP 4. Here is what we get after applying the branching procedure six times, …

STEP 5. … and this is the outcome after branching 12 times.

In the illustration accompanying Step 5, the length of the smallest branches is already smaller than their thickness. Continuing the procedure will not affect anything that we could detect with naked eye. So, we could as well imagine at this point that what we see in the last picture is the outcome of our branching operation applied repeatedly *infinitely* many times. Herein lies the point: after applying the branching operation infinitely many times we get a *fractal tree*, where each branch is similar to the whole tree. For example, the similarity taking the whole tree, to say, the left-hand side main branch is a composition of a translation (taking the root-point R_T of the tree to the root-point R_B of the branch; see Step 5), followed by a rotation around R_B through 60°, followed by a central similarity centered at R_B and with stretching factor 1/2. So, this fractal is a complete fractal.

The method we have used is a fairly standard way to construct a fractal: we start with an object and repeatedly apply to it a carefully chosen transformation. Repeatedly applying a single procedure is called iterating and the process is called **iteration**. After each step of iteration our object changes in such a way that the result is closer and closer to exhibiting self-similarity in an interesting way and thus becoming an *interesting* fractal. The full self-similarity is usually achieved at the limit of this process, that is, after iterating infinitely many times.

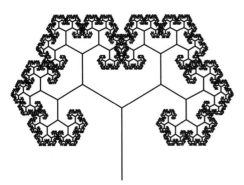

Since we are also interested in the visual aspect of fractals, we go a few steps further. First of all we notice that the tips of the branches of the last trees are far way from each other. So, we want to enlarge the ratio of 1:2 = 0.5 of the lengths of the baby branches with respect to their parent branches that we have used earlier. In Figure 3.3.2 we show what happens if we use a ratio of 0.65.

FIGURE 3.3.2 The ratio of the two consecutive branches in this fractal tree is 0.65.

We see that the branches became entangled due to their excessive length. The following question is then natural: what should the ratio of the lengths of the new branches with respect to the lengths of their parent branches be, so that the tree fractal we get has tips of the branches exactly touching?

The answer is: $\frac{1}{\text{Golden Ratio}}$! Justifying this answer requires basic knowledge of geometric series. In any case, we again see the golden ratio in action. The outcome is shown in Figure 3.3.3.

FIGURE 3.3.3 The ratio of the lengths of the new branches with respect to parent branches is 1/Golden Ratio, or approximately 0.618. The tips of the branches just touch.

We describe below two more examples of tree fractals. In the first case, we start with a vertical segment, show the outcome after applying one iteration, and then show the fractal we get after iterating the initial step "infinitely" many times (of course, in the pictures we will cheat, and apply a *large* number of iterations—not infinitely many of them; however, we should keep in mind that the real fractal is obtained after iterating infinitely many times).

STEP 1. We get two more branches after one iteration. This is not sufficient to describe the general step, for it is not clear whether the parts of the original trunk (the vertical part) will also bear new branches.

STEP 2. The second step clears any ambiguity: each part—trunk or branch—produces two more branches as in the first step.

STEP 3. The fractal is the result of iterating infinitely many times. Of course, we do not perform infinitely many steps to get the above picture: only six steps suffice to get the idea of how the final fractal should look.

The same idea of producing tree fractals can be realized in three dimensions as well.

 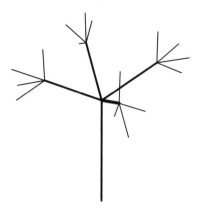

STEP 1. The initial vertical trunk produces four branches—keep in mind that this now happens in three dimensions so that what we see is just a perspective rendering.

STEP 2. The old branches produce new branches in the same manner.

STEP 3. A three-dimensional tree-fractal.

The final outcome (after iterating infinitely many times) is a three-dimensional tree fractal. We have mimicked nature using mathematical algorithms—not too successfully though, since one can rightfully argue that in nature the angles between the new and parent branches are never fixed to exactly 135°, as is the case in the tree shown in the illustrated Step 3 above. Nature involves a randomness that was not emulated here.

This is the right moment to touch on the so-called ***nondeterministic fractals***. Suppose we randomly choose the angles between the branches and, other than that, iterate using the same procedure as above (generating four new branches out of every old branch). Then we get nondeterministic fractal trees, a better rendering of natural trees. An example is given in Figure 3.3.4.

FIGURE 3.3.4 A nondeterministic three-dimensional fractal tree. ☐

We will now show a small gallery of various fractals. In the examples, we will first give the starting object, then the result after applying one or more iterations, ending with an approximation of the fractal.

Koch Snowflake

STEP 1. Start with a filled equilateral triangle.

STEP 2. Then add a smaller equilateral triangle over the middle third of each of the sides of the original triangle.

STEP 3. Continue the procedure: add a small equilateral triangle over the middle third of each of the segments in the previously obtained object.

STEP 4. Iterating infinitely many times we get a fractal called the **Koch snowflake**.

We will now show that the Koch snowflake is indeed an interesting fractal. In order to establish this, it suffices to specify a similarity that would move a large part M of the Koch snowflake within a proper part P of M. A description of such a similarity is given in Figure 3.3.5, where we also indicate the parts M and P of the Koch snowflake.

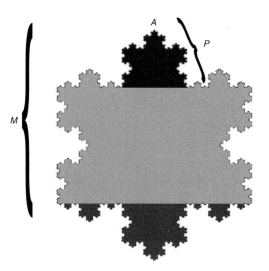

FIGURE 3.3.5 Showing that the Koch snowflake is a fractal.

The central similarity centered at the point A (see Figure 3.3.5) and with stretching factor 1/3 will map the part of the fractal denoted by M into a part of itself (denoted by P). There are infinitely many similarities sending a part of the fractal into a proper part of itself. For example, any central similarity centered at A and of stretching factor $1/3^n$, where n is any positive integer, will move M within a proper part of itself. □

The Dragon

The starting object is a (horizontal) line segment. Denote the end vertices of this segment by A and B, respectively. These two points will not change their positions throughout the whole procedure.

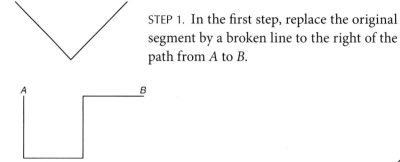

STEP 1. In the first step, replace the original segment by a broken line to the right of the path from A to B.

STEP 2. The first line segment is then replaced by two line segments to the *right* of the path from A to B after Step 1, while the second line segment is replaced by two line segments to the *left* of the path from A to B after Step 1. We alternate *right* and *left* in all of the iterations.

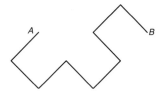

STEP 3. The result after performing one more step.

Can yo draw the next step in the procedure?

STEP 4. After infinitely many steps we get a fractal called the ***Dragon fractal***. The red-colored part is a part of the fractal that is self-similar. Can you describe a similarity that maps the points of that part of the dragon fractal onto a proper part of itself? □

Pentaflake

STEP 1. Start with a regular pentagon.

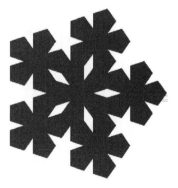

STEP 2. Then cut out triangular portions as shown above, getting as a result six small regular pentagons.

STEP 3. Iterate one more step.

STEP 4. After performing infinitely many iterations we get a fractal called a **Pentaflake**. Can you show that this is indeed an interesting fractal by describing a similarity that sends a large part of the pentaflake within a proper part of itself?

A Bit of Math. Fractal Dimension

A cursory look at the boundary of the Koch snowflake is all that is needed to notice that it possesses some fatness—that it is not exactly as 1-D as ordinary lines are. Neither does it appear that it is 2-D, as is, say, the interior of a filled square. So then, what is the dimension of the boundary of the Koch snowflake? In order to answer this question, we need a suitable definition of *dimension*. The definition given below is a watered-down version of the so-called Hausdorff dimension. Hausdorff was a mathematician who lived during the second half of the nineteenth century and the first half of the twentieth century. His exact definition of dimension is within the realm of *topology*, an important part of mathematics, which we will briefly consider in the last chapter.

Consider a square. We know it is 2-D. Our initial goal is to identify two-dimensionality of the square in a more definitive manner. To get there, start with the square and double it (i.e., double its edge length). Then the number of original squares that can fit the larger square increases four times. If instead of doubling we triple the size of the original square, then the number of original squares that can fit the tripled square increases by a factor of nine. In the first case, we compute the number:

$$\frac{\log \text{ (the factor by which the number of original squares has increased)}}{\log \text{ (the factor by which the size of the original square has increased)}}$$

(We use $\log c$ as short for $\log_{10} c$, and recall that $\log_{10} c = b$ means $10^b = c$.) Using standard mathematical notation, denoting the factor by which the number of original squares has increased by f, and denoting the factor by which the size of the square has increased by g, the number we have defined is $\dfrac{\log f}{\log g}$.

In case of doubling the square, this expression becomes $\dfrac{\log 4}{\log 2}$, which happens to be exactly 2. In case of tripling, the same formula gives $\dfrac{\log 9}{\log 2}$, which is also 2. So, we do get the dimension of the square using the formula $\dfrac{\log f}{\log g}$.

We now apply the same idea to the boundary of the Koch snowflake. Let us forget the fuzziness of that boundary and pretend that the Koch snowflake is of the rough shape depicted in Figure 3.3.6, as if our resolution is too rough to catch the details of the shape of the boundary. Using the length of the line segments in this picture as unit length we compute the length of the Koch snowflake to be 12. Now we triple the size of the snowflake keeping the same resolution—so, we can see as much of the details as in the above picture (but this time the snowflake is three times larger). We get the snowflake shown in Figure 3.3.7. We count the number of unit line segments and get 48. The factor of increase is 48/12 = 4. Using the above formula for (Hausdorff) dimension, we get $\dfrac{\log 4}{\log 3}$, which is approximately 1.2618. So, the dimension of the boundary of the Koch curve is NOT an integer. □

FIGURE 3.3.6

FIGURE 3.3.7

The fact that (some) fractals are of noninteger dimensions is of significance not only in mathematics but also in some other seemingly unrelated sciences. For example, it has been observed by analyzing the metabolism of many living organisms that they behave in many ways like fractals with noninteger dimensions.

Sierpinski Triangle

STEP 1. Start with a filled equilateral triangle and then remove from it the middle small equilateral triangle as shown. We are left with three small equilateral triangles.

STEP 2. Perform the same procedure to these three triangles.

STEP 3. Iterate the procedure infinitely many times; the outcome is a fractal commonly known by the name of **Sierpinski triangle** (also called **Sierpinski gasket**).

Every Sierpinski triangle is a complete fractal, for each of them is similar to a proper part of itself (Exercise 3). □

Flowers and Cyber-Flowers Again

FIGURE 3.3.8 The plant is almost self-similar with respect to a spiral similarity.

We start with a beautiful plant (Figure 3.3.8).

Returning back to the cyber-flower depicted in Figures 1.4.12, 3.1.5, and in Figure 3.1.6, we notice that, as is the case with the flower depicted in Figure 3.3.8, it is also *almost* self-similar with respect to the spiral similarity h^* described in Section 3.1, in the text preceding Figure 3.1.5. This spiral similarity sends the first point (or seed), the one that is closest to the center, to the second point, then the second point to the third point, and so on. The problem arises with the last point, for h^* sends it out of the cyber-flower. However, it is not hard to see how we could rectify the glitch: we simply apply h^* *infinitely many times* to the starting point so that we get a cyber-flower with infinitely many seeds, and so that there is no *last seed*. In this case, we

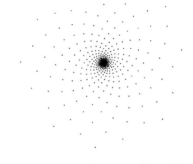

FIGURE 3.3.9 The spiral similarity h^* (described in Section 3.1) applied to a point a large number of times.

get that h^* indeed sends to whole cyber-flower into a proper part of itself. So Figure 3.3.9 represents only a small part of an infinite, unbounded, self-similar cyber-flower.

Exercises:

1. The Cantor set is obtained by deleting the middle open line segment out of a starting line segment, and then iterating the procedure on the remaining line segment infinitely many times. A line segment is open if the end points are excluded. The first three steps of the construction of the Cantor set is illustrated in Figure 3.3.10, where we start with a line segment of length 1.

   a. Why is the Cantor set a complete fractal? Find the center and the stretching factor of a similarity that moves the Cantor set within a proper part of itself.

   b. There are infinitely many similarities that send the Cantor set within a proper part of itself. Can you find them?

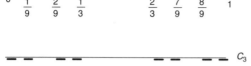

FIGURE 3.3.10 The first three steps of the construction of the Cantor set. The full construction is complete after *infinitely* many iterations.

2. We show below the steps in the construction of a fractal (the star fractal).

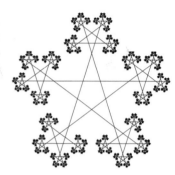

STEP 1. Star fractal: the construction after step 1.

STEP 2. Star fractal: the construction after step 2.

STEP 3. Star fractal final: infinitely many steps.

The last picture (in Step 3) shows what we get after six iterations. Since we cannot see the difference anyway, we will assume that this picture represents the fractal obtained by performing infinitely many iterations of the type indicated above.

  a. Draw at least one-fifth of the construction of the star fractal after three steps.

  b. Identify a maximal part M of the fractal in the illustration accompanying Step 3 that is self-similar to a proper part P of itself.

  c. Describe a similarity that sends the points of M onto the points of P.

3. Describe infinitely many similarities sending the whole Sierpinski triangle onto its proper part.

4. In the following two sets of illustrations (Figures 3.3.11 and 3.3.12), we give the trees we get after applying the first two iterations. Carefully consider the given iterations and then draw the tree obtained after applying one more step.

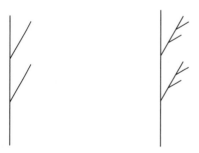

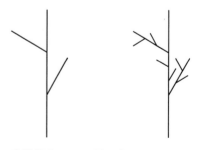

FIGURE 3.3.11 The first two steps of a tree-fractal.

FIGURE 3.3.12 The first two steps of another tree-fractal.

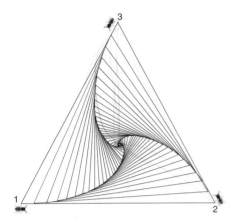

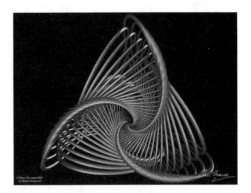

FIGURE 3.3.13 The triple spiral is the trajectory of the three pursuing bugs. Compare with Figures 3.2.5 and 3.2.6 of Section 3.2.

FIGURE 3.3.14 Marc Thomson, *Triospiro 2*, 2003.

5. In Figure 3.3.13, we show a fractal obtained from a (so-called) pursuing curve (An artwork on the same theme is shown in Figure 3.3.14.) Three bugs are initially positioned at the corners of an equilateral triangle. Bug 1 is pursuing bug 2, bug 2 is pursuing bug 3, and bug 3 is pursuing bug 1. All have the same speed. At every moment, each of the bugs directs its movement straight toward the pursued counterpart. The triangles you see are obtained by connecting the positions of the bugs at different moments. The trajectories of the three bugs make a triple-spiral fractal. Use the same idea to construct an approximation of the quadruple-spiral fractal obtained from the pursuing curve we get by starting with four bugs positioned at the corners of a square.
6. a. Show that there is no tiling with finitely many types of tiles and also is a complete fractal.
   b. Describe a tiling of the plane that is a complete fractal.
7. Find a complete fractal O such that for every positive number $\alpha < 1$, there is a similarity f with a stretching factor α such that $f(O)$ is a proper part of O.

3.4 JULIA SETS

We start with Figure 3.4.1. Then we zoom into the top (right) part of the coral-like object to get the insert shown in Figure 3.4.2.

Finally, we zoom in once again to the top (left) portion of the latter picture to get Figure 3.4.3.

There is no cheating: Figures 3.4.2 and 3.4.3 really show two different parts of the *coral*, and the last picture is just a magnified portion of the previous image of the same object. The self-similarity is evident—so what we see is a nice-looking fractal.

How are these fractals obtained and who discovered them? We start with the last question. As one might suspect, the pictures of the fractal shown above are computer generated. However, the underlying mathematical idea was discovered long before the age of

FIGURE 3.4.2

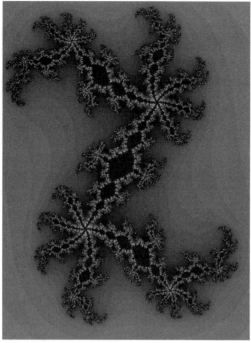

FIGURE 3.4.1

FIGURE 3.4.3

computers. In 1918, a 25-year-old mathematician named Gaston Julia (Figure 3.4.4) published a masterpiece of 199 pages (*Memoire sur l'iteration des fonctions rationelles*) where he inaugurated the study of *dynamic systems*.

His work was well ahead of his time, and it was mainly forgotten until it was revived in the 1980s. The fractals shown in this section are now called Julia sets in his honor.

Our main goal in this section is to describe the procedure that generates Julia sets. Once we accomplish this, we will show other fractals and identify their origins as well. More precise description of the examples is given by the end of this section, where we need the basic concept of complex numbers covered in Section 3.2.

FIGURE 3.4.4 Gaston Julia (1893–1978). He lost his nose during World War I.

Generating a Julia Set

We first provide the general description of the Julia sets, and then we follow it with examples. We need the notion of a bounded set of points in the plane: a set of points in the plane is **bounded** if all these points are within a (possibly very large) circle. Otherwise, we say that the set of points is **unbounded**. For example, the set of points in any square is bounded since any square can be encompassed within a circle. On the other hand, the points on any line are not bounded since there is no circle that will enclose the entire line.

Now we consider any transformation f of the points of the plane. Given any point A, we may apply f to it to get the image point $f(A)$. Then we can apply f to the new point $f(A)$ to get the point $f(f(A))$. Iterating this procedure *ad infinitum* yields the following infinite sequence of points in the plane: $A, f(A), f(f(A)), f(f(f(A))), ...$ If the set of points $\{A, f(A), f(f(A)), f(f(f(A))), ...\}$ is bounded, then we say that the starting point A belongs to the **prisoner set** of the transformation f. Thus, the prisoner set of f consists of all points A for which the set of points $A, f(A), f(f(A)), f(f(f(A))), ...$ is bounded. The other points of the plane constitute the **escape set** of f. That is, the escape set of f consists of those points A in the plane for which the sequence of points is unbounded. The boundary between these two sets is the **Julia set**. The prisoner set is sometimes called the **filled Julia set.** ☐

Example 1. The Julia Set of a Rotation

Suppose f is a rotation around a point C through an angle α. Then, for any point A in the plane, the points $A, f(A), f(f(A)), f(f(f(A))), ...$ are all on a circle centered at the point C and so are within any circle centered at the same point and with a larger radius. Consequently, this set is always bounded. We conclude that every point A on the plane belongs to the prisoner set of f. As a consequence the associated Julia set has no points at all. ☐

The last example shows that rotations do not produce interesting Julia sets. What about a symmetry f that is a rotation followed by a translation? In this case, we start with a point, and then we repeatedly apply the compositions of a rotation (say, around the origin) through a fixed angle, followed by a translation through a fixed vector. Could the set $\{A, f(A), f(f(A)), f(f(f(A))), ...\}$ be unbounded for some points A? It can be shown that this is not possible. In fact, somewhat surprisingly, even in this case all the points $\{A, f(A), f(f(A)), f(f(f(A))), ...\}$ lie on the same circle.* This shows that even in this case the prisoner set of points is the whole plane, and thus the Julia set is again empty. Indeed, symmetries (and similarities too) do not produce interesting Julia sets. We need to use more complicated transformations f to get interesting examples of the sort shown at the beginning of this section.

Example 2. A Simple Transformation with a Nontrivial Julia Set

Fix a point A in the plane and define a transformation f as follows: $f(A) = A$ and for every other point f is the same as translation through a fixed nonzero vector. The prisoner set and the Julia set for this transformation coincide and are made of infinitely many points (see Figure 3.4.5 and its caption). ☐

Example 3. The Squaring Transformation Composed with a Translation

Recall the transformation in Example (b) in Section 3.1, which we called the squaring transformation composed with a translation. In brief, we move points in the plane according to

---

* Moreover, for any two points A and B, the set of points $\{A, f(A), f(f(A)), f(f(f(A))), ...\}$ and $\{B, f(B), f(f(B)), f(f(f(B))), ...\}$ lie on two concentric circles.

FIGURE 3.4.5 The Julia set and the prisoner set consist of the points we indicate here (the sequence of points extends to the left without end). The points $A_1, A_2, A_3, \ldots$ end up at position A after repeated usage of the translation through the shown vector, and then they stay at A all the time.

the following rule (see the corresponding picture in Example (b) in Section 3.1): fix a line segment OX and call the point O the origin; for any other point A double the angle XOA; choose a point $s(A)$ that is OA^2 units away from O; translate along the given vector c. We denote the corresponding transformation by t.

We will now take a look at a specific transformation of this sort, taking the vector c to be the vector that starts at the origin O and ends at the point that is 0.2825 units to the right of the origin and 0.47375 units above OX. We note in passing that these two numbers are not chosen randomly; they are related to the first three pictures in this section. This is clarified in the next few paragraphs.

In order to see if a point A belongs to the prisoner set of t, we need to explore the set of points, $A, t(A), t(t(A)), t(t(t(A))), \ldots$, that we get by repeatedly applying t. We choose a specific point A that is 0.468 units to the right of O in the direction of X, and then 0.745 below OX. It is depicted in Figure 3.4.6. In Figure 3.4.7, we show both A and $t(A)$, whereas in Figure 3.4.8 we show the first 1201 points $A, t(A), t(t(A)), t(t(t(A))), \ldots \overbrace{t(t(\ldots t(A) \ldots))}^{1200}$ we get by repeatedly applying t starting with A.

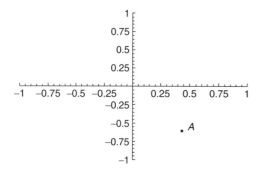

FIGURE 3.4.6 The starting point A. The horizontal line is the line through OX, and the vertical line is shown here only to indicate the vertical distances. The point O is at the intersection of the two lines.

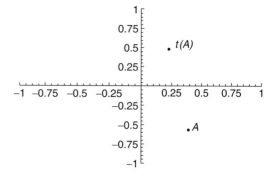

FIGURE 3.4.7 We show both the starting point A and the image point $t(A)$.

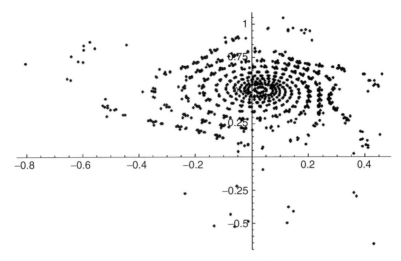

FIGURE 3.4.8 The points after applying transformation, t, up to 1200 times. As is visible, these points do not wander far away.

Apart from noticing an interesting pattern, we observe that all of the 1201 points stay very close to the point O (which is in the intersection of the two lines we show), and so, there is a chance that this point generates a bounded infinite sequence A, $t(A)$, $t(t(A))$, $t(t(t(A)))$, ... However, if we change the position of the starting point A by a bit, say by changing 0.468 (the horizontal distance to the starting point) to 0.4681 (keeping the vertical distance of -0.745 unchanged), then the sequence A, $t(A)$, $t(t(A))$, $t(t(t(A)))$, ... will not be bounded. Specifically, after *only* 450 iterations of t over the starting point, we get a point that is more than 10^{18924} units away from the origin O, an extremely large distance indeed.[*]

We can now identify the origin of the first three pictures in this section (Figures 3.4.1 through 3.4.3): they are all approximations of the prisoner set (and so, the Julia set too) of exactly the transformation t that has been just described. □

Varying the small vector c by just a bit (while keeping the same definition of the transformation t) may often profoundly affect the associated Julia set. In many of the cases, we get very wild fractals of stunning visual beauty. We will now show some (parts of) Julia sets, and in each case, we will describe the transformation t (by specifying the translation vector c). As in the above example, we will identify c by giving the horizontal and the vertical displacement (the coordinates). For example, $c = (1, 2)$ means that the vector c points in the direction 1 unit to the right and 2 units upward, whereas $(-3, 0.5)$ is a vector pointing 3 units to the left and 0.5 units up. Since small changes of the vector c sometimes affect violent changes of the associated Julia sets, we give coordinates of the vectors c with fine precision.

A brief explanation regarding the colors of the fractals is in order. As noted above, the colors are also produced algorithmically—they are not only for the visual effect, though one cannot deny their value in that respect. Most of the time the color code indicates how far a point is from being in the prisoner or escape set. For example, points in black color are

---

[*] According to some estimates, the diameter of the *observable* universe is less than 10^{40} cm long!

usually the points in the prisoner set, whereas the gradients in the escape set indicate how many iterations are needed so that theimage of the original point is sent at a more than (in advance) prescribed distance from the point O. We will occasionally use other coloring codes to put more emphasis on the visual effects (Figures 3.4.9 through 3.4.11).

FIGURE 3.4.9 $c = (0.4961718, -0.24974)$.

FIGURE 3.4.10 **(See color insert following page 144)** $c = (-0.1525, -0.65)$.

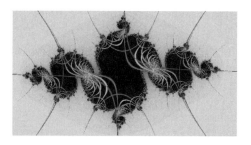

FIGURE 3.4.11 $c = (-0.78666, -0.09333)$.

More Complicated Examples (Optional; This Part Depends on Section 3.2)

As we have learned in Section 3.2, the "squaring transformation followed by a translation" is the same as the complex mapping $f(z) = z^2 + c$, moving a point corresponding to a complex number z to the point corresponding to the complex number $z^2 + c$. The associated prisoner set then consists of all points in the plane corresponding to the complex numbers z for which the infinite sequence $z, f(z), f(f(z)), f(f(f(z))), \ldots$ is bounded. There is no reason whatsoever that we have to consider only such functions f. For example, we could look at the Julia sets (and the prisoner sets) for a rational complex function, as are, for example, $\dfrac{z^3}{z^2 + 1} + 0.41333 + 0.61333i$ and $z - \dfrac{z^3 - 1}{3z^2}$. In Figures 3.4.12 and 3.4.13, we show approximations of the Julia set for these two functions.

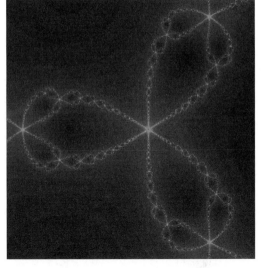

FIGURE 3.4.12 **(See color insert following page 144)** $f(z) = \dfrac{z^3}{z^2 + 1} + 0.41333 + 0.61333i.$

FIGURE 3.4.13 $f(z) = z - \dfrac{z^3 - 1}{3z^2}$

FIGURE 3.4.14 and Animation 3.4.1 $f(z) = \dfrac{1}{z^2}.$

FIGURE 3.4.15 and Animation 3.4.2.
So, naturalists observe, a flea has smaller fleas
that on him prey; and these have smaller still
to bite 'em; and so proceed *ad infinitum*.
Jonathan Swift (1667–1745) *Poetry, a Rhapsody*.

In Figure 3.4.14 (and Animation 3.4.1), we show (an approximation of) the Julia set of
the transformation $1/z^2$ that is similar to the transformation $1/z$ that we have encountered
in Sections 3.1 and 3.2. In Figure 3.4.15 (Animation 3.4.2), we show the Julia set associated
with a more complicated transformation defined as follows:

$$f(z) = \begin{cases} z^2 - 0.1875 - 0.765625i & \text{if Re}(z) < \text{Im}(z) \\ z^2 - 0.1875 - 0.765625i & \text{if Re}(z) \geq \text{Im}(z) \end{cases}$$

We end this section with yet another look at the Sierpinski triangle (see Section 3.3): it is the prisoner set of the transformation that moves every point *A* by means of a central similarity of a stretching factor 2 and (here is the twist) centered at that vertex of the triangle that is closest to the point *A*. If two or three vertices are equidistant from *A*, then we choose any of them as the center of the similarity. So, this transformation combines three central similarities with centers at the vertices of the starting triangle. A careful, but not difficult analysis, shows that the Julia set of this transformation is indeed the Sierpinski triangle.

Exercises:
1. What is the prisoner set of a translation? What about a reflection?
2. Identify the prisoner sets associated with central similarities of various stretching factors.
3. Fix a point *C* in the plane and consider the transformation *f* defined as follows: for any point *A*, *f(A)* is on the semiline from *C* to *A* and its distance to *C* is the square of the distance from *C* to *A*. Find the prisoner set of the transformation *f*.
4. The points in or on the circle A are not moved, and the points outside that circle are translated through the shown vector **v** (see Figure 3.4.16). Draw the Julia set of this transformation. [*Note*: the Julia set in this case will be unbounded.]
5. Consider the object A depicted in Figure 3.4.17 (the dotted line is just to see more clearly the geometry of A), and consider the following transformation *f*. The points within A as well as on the boundary of A are rotated 60° around the center O, whereas the points outside of A are translated through the shown vector **v**. Draw the Julia set for *f*.

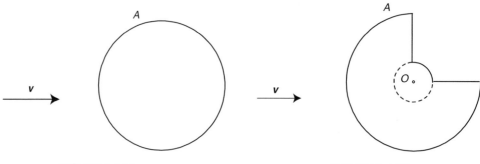

FIGURE 3.4.16   FIGURE 3.4.17

6. a. Starting with two points *A* and *B* on a line, we play an unusual game of ping-pong as follows: The players are positioned at the points *A* and *B*, and the player closer to the ping-ping ball sends it three times farther from himself (in the case when the ball is in the middle of *A* and *B*, the ball is hit by any of the two players). Each time the first player who hits the ball chooses the starting position of the ball on the line segment between the points *A* and *B*, the starting player wins if the ball stays between *A* and *B* after every hit by either of the two players. First experiment with a few starting positions for the ping-pong ball, then find the set of winning positions. [*Hint*: consider the Cantor set defined in Section 3.3.]

   b. Now start with three points *A*, *B*, and *C* in the plane, and three players positioned at these points. The rules are the same as above: the player closest to the ball hits it three times farther from his position (while it does not matter who does this if the ball is at equal distance between two or three players). The first player to hit the ball wins only if the ball always stays within the triangle with vertices *A*, *B*, and *C*. What is the set of winning positions this time?

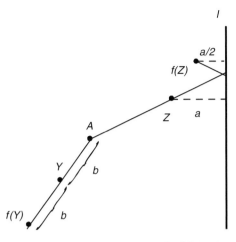

FIGURE 3.4.18

7. Start with two intersecting lines and define a transformation of the plane as follows: for any point on the plane, identify the line that is closer to it (or choose any of the two lines if the two distances are equal), and double the distance to that line along a perpendicular line passing through the point. Find the prisoner set of this transformation.

8. Fix a vertical line *l* and two points *A* and *B* out of that line, one to the left, and the other to the right of it (see Figure 3.4.18). We define a transformation *f* as follows:

   • $f(A) = A$ and $f(B) = B$; for every point X on the line *l*, $f(X) = X$.

   • For every point *Y* to the left of the line *l*, $Y \neq A$, we draw a ray starting at *A* and passing through *Y*. If that ray does not intersect *l*, then $f(Y)$ is the result of applying the central similarity centered at *A* and with stretching factor 2. If *Z* is a point to the left of the line *l* such that the ray from *A* to *Z* intersects the line *l*, then we bounce the point *Z* off the line *l* to a twice shorter distance (see Figure 3.4.18).

   • The action of transformation *f* over the points to the right of *l* is defined symmetrically. Find the prisoner set and the Julia set of *f*. [*Hint*: you might need to recall from high school that the geometric progression $1 + \frac{1}{2} + \frac{1}{2^2} + \frac{1}{2^3} + \frac{1}{2^4} + \frac{1}{2^5} + \cdots$ determines a number (which happens to be 2).]

3.5   CELLULAR AUTOMATA

Example 1

Start with a row of squares, some white, some black: we call that row ***the seed***. Then form the second row of squares right below the first row, and color the squares in the second row black or white depending only on the adjacent squares in the first row and according to the following rule.

Rule 18 (the numbering will be explained in a few paragraphs): for every square *S* in the second row, denote the three adjacent squares in the first row by *A*, *B*, and *C* in that order (so, the square *B* is directly above *S* and they share a common edge). Color the square *S* black if the square *B* is white, and exactly one of the squares *A* or *C* is black. Otherwise, color the square *S* white.

A visual description of Rule 18 is given in Figure 3.5.1.

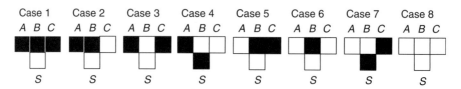

FIGURE 3.5.1 We shows all of the cases covered by Rule 18: as we see, the square S should be colored black only in two of the possible eight configurations of the colors of the squares A, B, and C.

Suppose our starting row of squares contains only one black square. We apply Rule 18 to that row to get the second row of squares, then we repeat the same procedure (we iterate) two more times to get the resulting rows. The resulting rows (the parts containing the black squares) are shown in Figure 3.5.2.

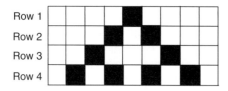

FIGURE 3.5.2 Apply Rule 18 to the first row to get the second row of squares, then apply it to the second row to get the third row, and finally, apply it once again to get the last (fourth) row.

FIGURE 3.5.3

Now imagine that our rows of squares are unbounded on both sides. Suppose also that we iterate Rule 18 infinitely many times so that we get infinitely many rows. What do we get?

Rule 18 applied infinitely many times, starting with a row containing a single black square, produces an object that is in many ways similar to the Sierpinski triangle. In Figure 3.5.3 we show the first 512 rows. Since there are infinitely many rows in the final design, the pattern should extend downward without end. In a way, we get the Sierpinski triangle from within. An artistic perspective on this example is shown in Figure 3.5.4.

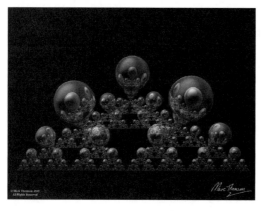

FIGURE 3.5.4 Marc Thomson, Reflecting Metallic Spheres, 2003. □

In the above example (Rule 18), we have described one particular 1-D cellular automation. In general (and informally), a 1-D cellular automation is a rule that is iterated on a row of objects (usually forming a frieze pattern) in various *states*, so that in each step, the state of every object in the row changes or not, depending *uniformly* on the state of that object, and on the states of the two neighboring objects in the row. So, after one application of a 1-D automation to a row of objects in various states, we get a new row of the same objects, but with new states of the objects.

In this context, the objects in the rows are usually referred to as cells, and they are usually represented by squares. The word *uniformly* means that there is a single rule that is consistently applied to all objects in the rows. We will indicate the states by coloring the objects. In Example 1 (this section), there were only two states, and so we used two colors (black and white).

If there are two states, then all eight of the possibilities of the states of a cell and its two neighboring cells are given in the top row of Figure 3.5.1 (the states *A*, *B*, and *C*). Then, each sequence of eight choices for the color (black or white) of the cell in the second row gives one rule or one one-dimensional cellular automation. One such sequence of eight choices gave rise to the rule that we have called Rule 18. Here, for example, is one more such sequence of eight choices.

Example 2

This rule, Rule 86 (Figure 3.5.5)—the notation will be explained later in this example—generates a sequence of rows as shown in Figures 3.5.6 and 3.5.7; the seed is an infinite row containing only one black square (and all the other squares are white).

In general, for each of the eight cases of configurations of states around a square (states *A*, *B*, and *C*), we have two choices for *S*: black and white. So, there are a total of $2^8 = 256$ rules, and hence there are 256 1-D cellular automata.

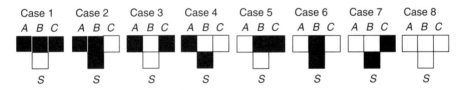

FIGURE 3.5.5 **Rule 86.**

FIGURE 3.5.6 We apply Rule 86 47 times. The starting row (top row) has only one black square.

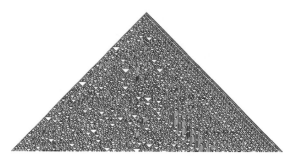

FIGURE 3.5.7 Here are the first 512 rows obtained by applying Rule 86 511 times starting with the top row. It is interesting that there is chaos at the left-hand side part, and the relative order at the right-hand side part of the output. ☐

Digression—Regarding Notation

Here is how the numbering of the rules is chosen. In each of the eight states, choose the digit 1 if S is chosen to be black, and the digit 0 if S is chosen to be white. Then, each choice of the eight states corresponds to a sequence of eight digits containing 0 or 1. For example, Rule 18 corresponds to the sequence 00010010, and Rule 86 gives the sequence 01010110. Now, to each such sequence, there corresponds the number (first digit) $\times 2^7$ + (second digit) $\times 2^6$ + ... + (seventh digit) $\times 2^1$ + (eighth digit) $\times 2^0$, and that number specifies the number of the rule. So, for example, the number 18 is obtained from the sequence 00010010 as follows: $18 = 0 \times 2^7 + 0 \times 2^6 + 0 \times 2^5 + 1 \times 2^4 + 0 \times 2^3 + 0 \times 2^2 + 1 \times 2^1 + 0 \times 2^0$, and the number 86 in Rule 86 comes from $86 = 0 \times 2^7 + 1 \times 2^6 + 0 \times 2^5 + 1 \times 2^4 + 0 \times 2^3 + 1 \times 2^2 + 1 \times 2^1 + 0 \times 2^0$. In passing we note that we have just described the *binary system* of writing numbers using the two digits 0 and 1. So, summarizing one of the above examples in terms of this new notion, the number 10010 is the binary notation for the number 18, meaning that $18 = \mathbf{1} \times 2^4 + \mathbf{0} \times 2^3 + \mathbf{0} \times 2^2 + \mathbf{1} \times 2^1 + \mathbf{0} \times 2^0$. ☐

Recall that a regular tiling of the plane is a tiling that uses regular polygons as tiles, and recall that the only such tilings consist of equilateral triangles, squares, or regular hexagons as tiles. A (restricted) two-dimensional cellular automation consists of a regular tiling of the plane where each of the tiles is colored black or white, together with a rule that tells us how to change the coloring of each of the tiles depending on the colors of the neighboring tiles. Initially, we consider two tiles in the regular tiling to be neighbors if they share a common edge. We start with an example.

Example 3. A Two-Dimensional Automation

We consider the regular tiling of the plane with squares. One cellular automation rule, tentatively called Rule X, is given in Figure 3.5.8. We list all possible black–white colorings of a square and its four neighbors. We start with all five squares being white (top-left corner), and, after carefully listing all combinations, we end up with all five squares being black (bottom-right corner). Below each configuration of cells (squares), we indicate with an arrow what happens with the middle square. For example, the arrow below the top-left configuration (all five squares are white) points to a white square, indicating that according to this rule if

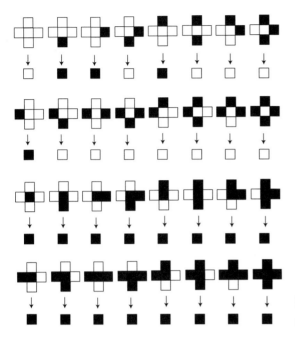

FIGURE 3.5.8 This picture specifies a two-dimensional cellular automation.

the middle square is white and if all of its neighbors are white, then the middle square stays white. The next arrow (second from the left in the top row) tells us that the middle square changes from white to black if the square straight below it is black, while the other three neighbors are all white. There are (as visible in Figure 3.5.8) 32 possible cases, with arrows indicating what happens with the middle square in each of the cases. The last 16 cases in this rule tell us that once a square is black, it will never turn into a white square. An automation of this type (with four neighbors, and with these four neighbors together with the middle cell influencing the life of the latter) will be called a *five-neighbor cellular automation*.

Suppose the initial tiling of the plane contains only one black square. In Steps 1, 2, and 3 below we see what happens in the neighborhood of that square after applying the above rule once, twice and thrice, respectively.

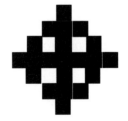

STEP 1. This is how the neighborhood of one black square changes after applying Rule *X* only once.

STEP 2. We have applied Rule *X* twice.

STEP 3. As we see, the neighborhood only after applying Rule *X* thrice is not so simple.

In Figure 3.5.9, we show what happens after 45 iterations of Rule *X*, starting with a tiling containing only one black square.

FIGURE 3.5.9 This is what we get after iterating Rule X 45 times, starting with a single black square.

Digression—On the Number of Cellular Automata

We will now explain why we used the generic name Rule X for the automation we have described above, rather than numbering it in the manner we have numbered the rules of one-dimensional automata. Recall that in case of one-dimensional automata there were eight possible cases of cell configurations, giving rise to $2^8 = 256$ rules. In case of cellular automata of the type considered in Example 3, there are 32 types of colored configurations in the neighborhood of a cell, so that there are $2^{32} = 4294967296$ five-neighbor cellular automata. Our Rule X happens to be Rule 1753284607 in the ordered list of rules. So, we were reluctant to use such a large number as the name of a cellular automation.

N	N	N
N	M	N
N	N	N

FIGURE 3.5.10

Continuing on the same subject, it is relatively easy to see that there are 512 configurations of colored neighbors of a square if we consider the neighbors of squares to be not only the four squares that share a common edge, but also the four squares that share a common corner vertex (see Figure 3.5.10, where the middle square is denoted by M, while the neighboring squares are all labeled by N). Cellular automata in which the state of each cell M is influenced by the states of the cells N, as shown in Figure 3.5.10, will be called here nine-neighbor cellular automata. It follows by an argument similar to the one we have used to count the number of one-dimensional cellular automata and the number of five-neighbor two-dimensional cellular automata that there are

$$2^{512} = 1340780792994259709957402499820584612747365820592393377723561443721764$$
$$0300735469768018742981669034276900318581864860508537538828119465699464$$
$$33649006084096$$

many nine-neighbor cellular automata. How large is this number? Suppose you play with each and every such regular automation for only 1s. Then you would need about

4359071970564982931353394519287689258049627360523432095857899449815908509569
2711509057279631472714535509102743271711307654994677265223710

billions of years to complete your exploration of cellular automata of this sort. For comparison, the age of the so-called *observable universe* (a concept of rather vague meaning), starting with the assumed *Big–Bang*, is estimated to be of the order of 30 billion years. □

Some of the rules defining regular automata can be described clearly without listing explicitly all possible neighborhood configurations of black–white squares. For example, in the Game of Life, invented by John Convey, the emergence of black (or colored) squares is described as the birth of cells, whereas the change from a black square to a white square is referred to as the death of the associated cell. The colored cells are called living cells. In the Game of Life, the birth of a middle cell comes as a result of the presence of a certain (in advance given) number of living cells in its neighborhood, and so does its death. For example, in the Rule 123/45, the numbers 1, 2, and 3 indicate that a living (black) cell will survive if there are exactly 1, 2, or 3 living cells in the neighborhood (and it will die otherwise), whereas the numbers 4 and 5 indicate that a dead (white) cell will be (re)born if there are exactly 4 or 5 many living cells around it (and it will stay dead otherwise).

Example 4. A Game of Life

Here is, for example, what happens when we apply Rule 123/45 to a tiling that has one cluster of black cells arranged in a 5 × 5 square array (Figures 3.5.11a through 3.5.11c).

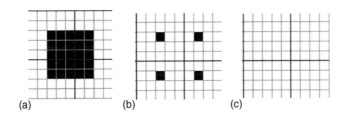

(a)　　　　(b)　　　　(c)

FIGURE 3.5.11 In the Game of Life, Rule 123/45 is applied to a 5 × 5 cluster of black (blue) cells. Only the corner cells survive after one application of the rule, and all the living cells are gone after the next iteration.

Rule 123/45 killed all the cells in the seed in a couple of steps. Let us experiment a bit with another rule, Rule 45/123: the living cells survive with a neighborhood of four or five living cells, whereas the new cells are born (turn from white to black) if there are one, two, or three living cells in their neighborhood (see Figures 3.5.12 and 3.5.13).

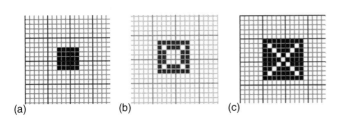

(a)　　　　(b)　　　　(c)

FIGURE 3.5.12 In the Game of Life, Rule 45/123 is applied to a 5 × 5 cluster of black (blue) cells. Some cells die, but new ones are born.

FIGURE 3.5.13 The population will spread. Here we see the result after 50 iterations of the rule. □

We have only considered very few rules out of the large number of nine-neighbor rules for cellular automata. There are other types of cellular automata. For example, we can have more states (rather than only two states considered in the above examples), we may set nonadjacent cells (those that are farther) to influence the states of the cells, or we may consider different types of patterns, regular (as the hexagonal tiling) or not. All these can be applied to infinitely many starting configurations to get infinitely many processes. Some of these generate interesting patterns (Figures 3.5.14 and 3.5.15).

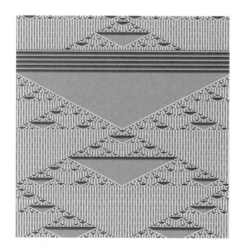

FIGURE 3.5.14 This is a one-dimensional automation applied on a random row of squares. The state of a cell in this rule depends on the states of the immediate and next to the immediate cells.

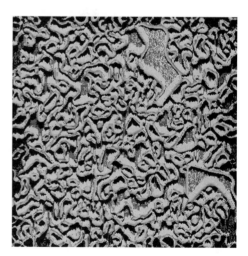

FIGURE 3.5.15 We show a random graphic generated by a cellular automation from the family of automata called Margolus (in which the positions of the cells change depending on the neighborhood configuration).

The concept of cellular automata was introduced in the 1950s by many people, some of them arriving at the idea through their research in the area of artificial intelligence and computers, the others getting there through purely mathematical investigations of infinite sequences of 0 and 1 (white and black).

Cellular automata are used as simplified simulators of various processes. Examples include crystallization simulation, gas behavior simulation, ecosystem modeling, immune systems modeling, interaction of particles in physics, modeling of self-reproducing machines in computer science, and many more. Cellular automata are also used as graphic self-generators—one of the reasons we have included them here. According to some people, soon we would not be able to watch television for an hour without seeing some kind of graphic generated by cellular automation. Aside from their applicability, the shear beauty of this simple concept has a value in itself.

Exercises:

1. Draw the next two rows if the one-dimensional automation defined in Example 1 (this section) is applied to the following starting row:

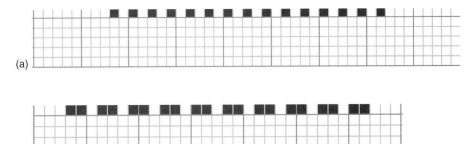

2. Which rule for a one-dimensional automation was used to obtain the following design, starting with the top row? Draw a picture of the rule according to Figure 3.5.1. [*Note:* there may be more than one solution.]

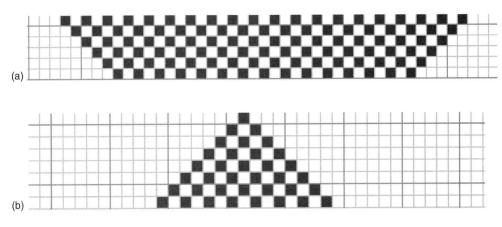

3. In the following illustration, we see a one-dimensional cellular automation with infinitely many states, where each state is denoted by a nonnegative integer. The rows, as was the case earlier, extend unboundedly on both sides. We only show the first five rows

(obtained by applying the automation four times, starting with an application on the first seed row). Note also that in this automation, each square in the rows starting from the second is adjacent to only two squares in the preceding row.

0	0	0	0	0	1	0	0	0	0	0
	0	0	0	0	1	1	0	0	0	0
0	0	0	0	1	2	1	0	0	0	0
	0	0	0	1	3	3	1	0	0	0
0	0	0	1	4	6	4	1	0	0	0

a. Identify the rule in this automation, and describe the next row.

b. Expand the binomial $(a + b)^n$ for various small integer values of n. How are the coefficients in the expansion (the numbers in front of the powers of a and b) related to the states in the above automation?

4. Apply Rule X (in Example 3 above) once to the following colored regular tiling of the plane.

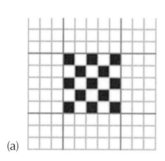

(a)

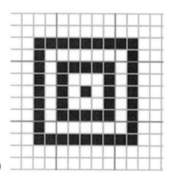

(b)

5. Describe a rule in the Game of Life that changes the pattern to the left into the pattern to the right.

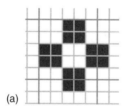

(a)

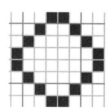

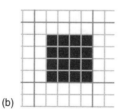

(b)

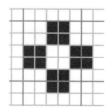

6. Start with a colored hexagonal tiling with only one black hexagon, and apply the following rule two times consecutively: if a living (black) cell has exactly one living (black) neighbor it survives, and it dies otherwise; if a dead (white) cell has exactly one living neighbor, then it is (re)born, and it stays dead otherwise.

7. a. Start with a square tiling of the plane with exactly one black square B. Find a two-dimensional cellular automation that changes (after a while) the starting colored tiling into a tiling with exactly one black square that is 10 squares to the right of the starting black square B.

    b. Start with any design of black and white squares in the regular tiling made of squares, and find a two-dimensional five-neighbor cellular automata that would shift the whole pattern 10 squares to the left after 10 applications.

8. Find a web page with a real-time script for nine-neighbor cellular automata, randomly choose a large number smaller than 2^{512}, play with the corresponding regular automation by applying it to various seeds, then call it your own. It is extremely unlikely that anyone has chosen or will ever choose the same rule.

CHAPTER 4

Hyperbolic Geometry

We will now take a look at the unusual world of non-Euclidean geometries, specifically at the *hyperbolic geometry*.

4.1 NON-EUCLIDEAN GEOMETRIES: BACKGROUND AND SOME HISTORY

Recall the five Euclidean postulates of the planar geometry (see Section 1.1), in particular, the fifth postulate that states the following:

> For every line *l* and every point *P* that does not lie on *l*, there exists a unique line *m* through *P* and parallel to *l* (see Figure 4.1.1).

For 2000 years many learned people tried to show that the fifth postulate could be deduced from the other four postulates (see Section 1.1), for it appeared that this should have been the case. Nobody succeeded, but the shear volume of unsuccessful attempts prompted other people to look in the opposite direction: perhaps the fifth postulate is independent from the other four postulates, and thus it is not deducible from them. Perhaps there exists a world, with another geometry, where the fifth postulate is not true.

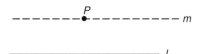

FIGURE 4.1.1 The fifth Euclidean postulate.

The mystery unraveled in the first part of the nineteenth century. Nikolai Lobachevsky (Figure 4.1.2), a Russian mathematician at the University of Kazan in southern Russia, published in 1829 a thorough account of a new geometry that he called "imaginary geometry." His paper (in Russian) was neglected until 1840 when he published a treatise in German. Gauss, arguably the most important mathematician of the nineteenth century, praised Lobachevsky's work,

FIGURE 4.1.2 Nikolai Lobachevsky (1792–1856).

although he claimed priority—apparently he did substantial work on the subject and antici-
pated the results (but never published them). The plot thickened with the appearance of yet
another mathematician who pursued the same idea independently: the Hungarian Janos
Bolyai published his work in an appendix of a book by his father in 1831. Upon receiving
a discouraging reply from Gauss (who told Bolyai that he had merely retraced Gauss' own
work from years ago) and upon learning that Lobachevsky had already published the same
discovery before him, Bolyai stopped doing mathematics. Lobachevsky's destiny was not
better: he was unappreciated and isolated during his lifetime, eventually even fired by the
University of Kazan, despite being an outstanding teacher, researcher, and administrator.
He died blind in 1856, a year after Gauss, and 4 years before Bolyai.

The new geometry used to be called Lobachevskian geometry; it is now called hyperbolic geometry, the term introduced by the German mathematician Felix Klein, whom
we will encounter again in Chapter 6. As we have already stated, in the new, hyperbolic
geometry, Euclid's fifth postulate is not true while the other four postulates are valid. Of
course, it was not sufficient just to change the fifth postulate, for the new geometry thus
obtained could have been self-contradictory, and so meaningless. The main problem was
to show that this was not the case, and the main route to accomplish that was to construct
an actual model (a structure) where all of the four Euclidean axioms together with the new
fifth postulate were true.

Here is the new, fifth postulate that defines the hyperbolic geometry on the plane.

The fifth Hyperbolic postulate. For every line *l* and every point *P* that does not lie on *l*, there
exist infinitely many lines through *P* that are parallel to *l* (see Figure 4.1.3).

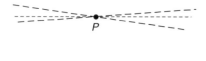

FIGURE 4.1.3 A point *P* out of the
line *l*, and many lines through *P*
parallel to the line *l*.

Before we take a closer look at a specific model of
hyperbolic geometry, we will consider an exotic world,
where some of our preconceived notions fail.

In Figure 4.1.4, we see a lonely Dadat in a strange
world with a checkered ground, a red sky, and a magic
crystal ball floating above the ground. The crystal ball
reflects the scene and we can spot the image of Dadat
at the bottom of the ball. For us, the *real* world is the
three-dimensional scene. There, we see the checkered
plane with two sets of mutually parallel lines, those receding from us represented in the
perspective drawing as lines intersecting at the horizon (we will say more about perspective drawings in Chapter 5).

Now pay attention to the crystal ball reflection (see Figure 4.1.5). Imagine for a moment
that the reflection on the crystal ball is in fact a world in itself, and that the image of Dadat
on the crystal ball is a real, living two-dimensional being, seeing (and perceiving) exactly
the same objects and in almost the same way that its three-dimensional counterpart sees
on the plane. Then our own perception of the two-dimensional world on the surface of
the ball is not the same as the perception of the two-dimensional Dadat living in that
world, or, we might say, what we see is not what really is. In Figure 4.1.5 (the view of the

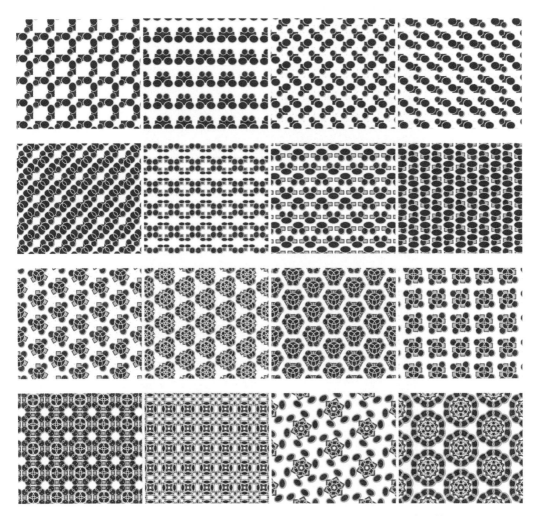

COLOR FIGURE 2.5.2 Wallpaper designs with the remaining 16 types of wallpaper groups.

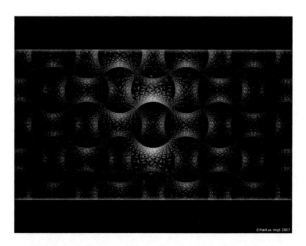

COLOR FIGURE 2.5.7 Marcus Vogt. *Tiling Shapes 06*, Apophysis fractal generator, then Photoshop, 2007.

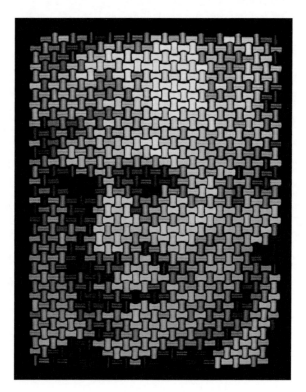

COLOR FIGURE 2.5.8 Ken Knowlton. *Aaron Feuerstein; Spools of Thread*, 32 in. × 26 in. Collection Aaron Feuerstein, © 2001 Ken Knowlton.

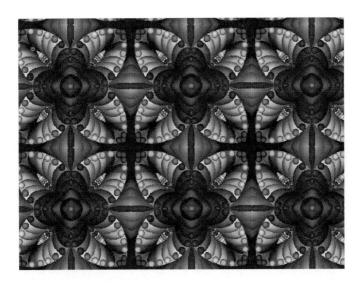

COLOR FIGURE 2.5.25 Floral tiling of the plane, type 4444.

COLOR FIGURE 2.6.10 John Osborn. *Bats*, ink and watercolor, 1990.

COLOR FIGURE 2.6.17 James J. Lemon: *voroscreen006*, 2006.

COLOR FIGURE 3.4.10 c = (−0.1525, −0.65).

COLOR FIGURE 3.4.12 $f(z) = \dfrac{z^3}{z^2 + 1} + 0.41333 + 0.61333i.$

COLOR FIGURE 5.1.12 *The Assumption of Virgin Mary,* fourteenth century, Ohrid, Macedonia.

COLOR FIGURE 5.1.14 Jean Fouquet, *The royal banquet*, detail from *Les grandes chroniques de France*, 1420–1480.

COLOR FIGURE 5.2.12 Dick Termes, *Pantheon*, 13″ sphere, 1998. Here is how Dick Termes describes the six-point perspective: "Six point perspective shows up when you study what happens to the lines that create a cubical room. When you stand inside that room you will notice there are three sets of parallel lines that create that room. If you look closely at those lines one set of those lines vanishes to the north and if you turn you will see that same set vanishes to the south. The next set vanishes off to the east and the same set vanishes off to the west. The last set of lines of that room project up directly above your head and the same lines project down below your feet. That is six point perspective. When you draw this on the sphere it is much more obvious."

COLOR FIGURE 5.4.4 (Art on a spherical canvas) Dick Termes, *Emptiness*, 24″ sphere, 1986.

COLOR FIGURE 5.5.3 Jos Leys, *Balinv126,* 2005. This is an approximation of the image of a set of spheres under inversion with respect to a sphere.

COLOR FIGURE 5.5.12 Marcus Vogt, *Tower Series 04*, XenoDream 3D fractal software, then Photoshop, 2007.

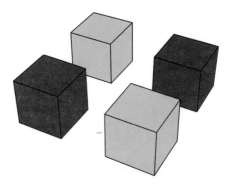

COLOR FIGURE 5.5.13 In the initial configuration we start with four colored cubes; we do not show the white cubes.

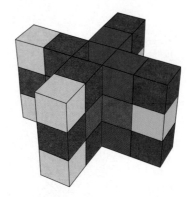

COLOR FIGURE 5.5.14 A cube becomes or stays colored if it is adjacent to exactly two colored cubes. This is what we get in the next step.

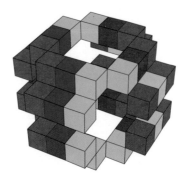

COLOR FIGURE 5.5.15 One more step: the population of the colored cubes spreads.

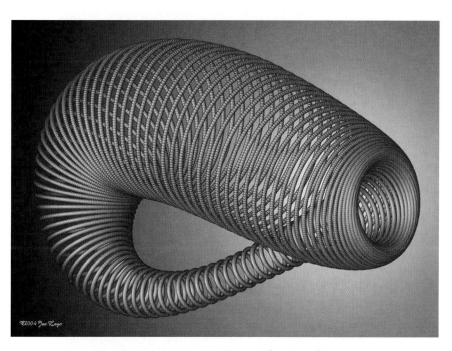

COLOR FIGURE 6.3.9 Jos Leys. *Klein Bottle*, 2004.

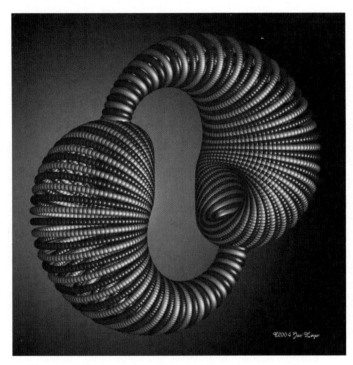

COLOR FIGURE 6.3.14 Jos Leys. *Bonan-Jeener's Klein Surface 1,* 2004.

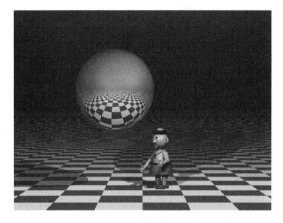

FIGURE 4.1.4 A crystal ball, and Dadat on a plane tiled with squares.

FIGURE 4.1.5 The world on the surface of the crystal ball.

crystal ball from straight below it), we see curved lines (the reflection of the checkered ground). However, for the people in this world on the surface of the crystal ball (and the two-dimensional Dadat in particular), these curved lines are just ordinary straight lines vanishing at their own horizon. More importantly, from our point of view, the world in the crystal ball is a small world within the limits of the bounded ball, and we see that the usual (Euclidean) sizes of the squares on the checkered board decrease as we go in the direction of the periphery of the image of the lower half of the ball. But for the people on the surface of the crystal ball, the horizon (the circular periphery) is as far from them as is our horizon from us—infinitely far—and for them the squares are just as *squarey* as the ones on the checkered plane below the ball. Our usual (Euclidean) distance *does not describe* faithfully the geometry of the crystal ball world. (In Figure 4.1.6, we show another crystal ball reflection.)

This example is not a model of the hyperbolic geometry—in fact it is a model of the usual Euclidean geometry (under natural assumptions that we will not get into). However, it does possess some unusual properties (from our point of view) and so we use it as a prelude to the world of hyperbolic geometry—and the Poincaré model of that geometry in particular—where the objects share some common features with the objects we see on the surface of the crystal ball.

Before we expand further on the theme of hyperbolic geometry, we note that there is one more type of planar geometry directly related to Euclid's fifth postulate and where the other four postulates are

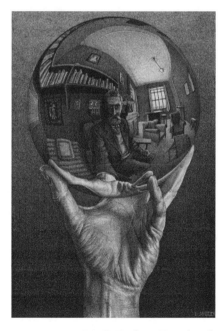

FIGURE 4.1.6 M. C. Escher. *Hand with Reflecting Globe*, lithograph, 1935.

valid. You will notice that the fifth postulate of the Euclidean geometry stipulates the existence of exactly one parallel line through a given point and to a given line, whereas the fifth postulate in the hyperbolic geometry allows infinitely many such lines. It can be shown that if the first four postulates are true, and if we accept that there are two or more lines passing through a point out of a fixed line and parallel to the fixed line, then there must be infinitely many such lines. There is one more possibility left: given a line and a point out of the line, there are *no* lines at all passing through the point and parallel to the given line. This gives rise to yet another form of geometry, called elliptic geometry. We will just touch on that subject and spend more time on the hyperbolic geometry (the latter being more interesting from an artistic point of view).

4.2 INVERSION

We have briefly encountered inversions in Section 3.1. Since we need them in the following sections, and since they are interesting (partial) transformations of the plane in their own right, we now describe them more thoroughly.

INVERSION 1. THE ACTION ON THE POINTS IN THE CIRCLE OF INVERSION

In Steps 1–3 illustrated below, we describe again how points within a circle are moved by inversion with respect to that circle.

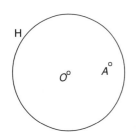 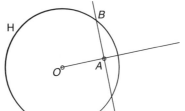 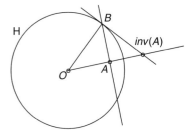

STEP 1. A circle H, its center *O* and a point *A*.

STEP 2. Join *O* and *A*, erect a perpendicular at *A*, and find the intersection point *B*.

STEP 3. Join *O* and *B* and draw the tangent line to the circle at *B* by erecting the perpendicular to *OB* at *B*. It intersects the line *OA* at *inv(A)*.

The closer the point *A* is to the center *O*, the farther the image *inv(A)* is from *O*. The point *O* (the center of H) is a special point, and the reason that this (or any other) circle inversion is a *partial* transformation is because it does not act on the center *O* at all. Loosely speaking, the center *O* is sent to *infinity* via the circular inversion. Conversely, no point in the plane will be moved to *O* under the inversion with respect to H. ☐

The points on the circle H are not moved via inversion with respect to H (they stay where they are).

INVERSION 2. THE ACTION ON THE POINTS OUT OF THE CIRCLE OF INVERSION

The points outside H are moved within H in the manner that reverses the procedure described in Steps 1–3 above. An explicit description is given in Steps 1–3 below. This construction relies on the following theorem of Euclidean geometry: an angle inscribed in a circle and over the diameter of the circle must be a right angle (Figure 4.2.1).

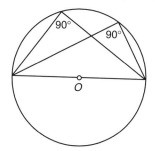

FIGURE 4.2.1 Angles over diameters are right angles.

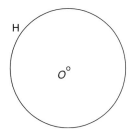
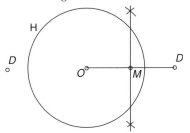
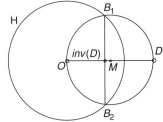

STEP 1. The point D is now outside the circle H.

STEP 2. Find the midpoint M of the segment OD.

STEP 3. The circle centered at M and passing through O and D intersects H at B_1 and B_2. The image $inv(D)$ of D is then on the intersection of OD and B_1B_2. □

We conclude that inversions with respect to a circle H shuffle points in the plane in a very special way: the points outside H move within H, and the points in H move out of H with one exception: the center of H is not within the scope of the inversion. The points on the circle H stay where they are.

Inversions in general do affect the shapes of planar objects. The following theorem is a partial description of how inversions change shapes. We fix the inversion circle H and continue to denote the associated inversion by inv.

Theorem:

a. If C is a circle that does not pass through the center of the inversion circle H, then its image $inv(C)$ is a circle.

b. If C is a circle that passes through the center of the inversion circle H, then its image $inv(C)$ is a line.

c. If L is a line that does not pass through the center of the inversion circle H, then its image $inv(L)$ is a circle with one missing point.

d. If L is a line that passes through the center of the inversion circle H, then its image $inv(L)$ is a line with one missing point (the center of H).

This theorem may be rephrased as follows: inversions circles and lines to move circles or lines. We illustrate a few of the possibilities in Figures 4.2.2 through 4.2.4.

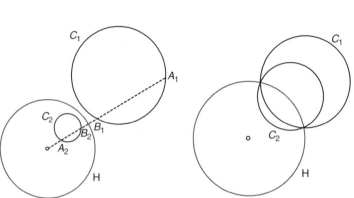

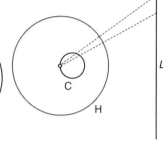

FIGURE 4.2.4 The inversion with respect to H moves the line L to the circle C

FIGURE 4.2.2 The inversion with respect to H moves C_1 to C_2 and vice versa.

FIGURE 4.2.3 In this case both C_1 and C_2 intersect H.

passing through the center of H, and it sends the circle C to the line L.

A Bit of Math (Optional; Needs Section 3.2)

We will give the idea of the proof of part (c). We will take a specific line (the vertical line $x = 2$), and a specific inversion circle (centered at the origin and of radius 1; we will denote it by H). The general proof of part (c) is not much more complicated than the proof we will see in this particular instance. This case is illustrated in Figure 4.2.4.

We will show that the image of any point on the vertical line $x = 2$ under the inversion with respect to the circle H is a point on the circle centered at the point $(1/4, 0)$ and of radius $1/4$. Take any point on the vertical line. Since its real part is 2, the complex number associated to any such point is of the form $2 + ai$, for some number a. According to Section 3.2, the image of that point under the inversion with respect to H is the point associated to the complex number $\dfrac{1}{2 - ai}$. To show that this is a point on the circle centered at $(1/4, 0)$ and of radius $1/4$, it suffices to show that the distance between the points associated to the complex numbers $\dfrac{1}{2 - ai}$ and $1/4$ (viewed as a complex number) is $1/4$. Referring to Section 3.2 again, this means that we need to show that $\left| \dfrac{1}{2 - ai} - \dfrac{1}{4} \right|$ is $1/4$. The computation follows

$\left(\text{in the first few steps we simplify the expression } \dfrac{1}{2 - ai} - \dfrac{1}{4} \right): \left| \dfrac{1}{2 - ai} - \dfrac{1}{4} \right| = \left| \dfrac{4 - (2 - ai)}{4(2 - ai)} \right| =$

$\dfrac{1}{4} \left| \dfrac{2 + ai}{2 - ai} \right| = \dfrac{1}{4} \dfrac{|2 + ai|}{|2 - ai|} = \dfrac{1}{4} \dfrac{\sqrt{4 + a^2}}{\sqrt{4 + a^2}} = \dfrac{1}{4}.$ $\square$

Here is what happens with a chessboard after we apply inversion with respect to a circle to it (Figures 4.2.5 and 4.2.6).

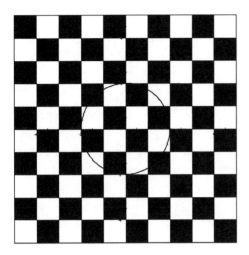

FIGURE 4.2.5 A 10×10 chessboard is inverted with respect to the shown circle.

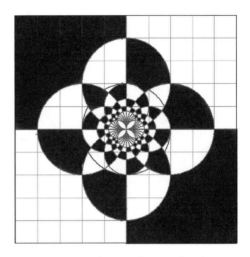

FIGURE 4.2.6 This is the result after we apply the inversion.

We do the same with the Sierpinski triangle (Figures 4.2.7 and 4.2.8): the outer triangle is such that each of its sides is 2 units long and the inversion circle is positioned at the center of the triangle and is of radius 1.

FIGURE 4.2.7 Six iterations of the Sierpinski triangle. The length of the side of the large triangle is 2. The circle of inversion is shown.

FIGURE 4.2.8 Inverting the Sierpinski triangle with respect to the circle of inversion (also shown).

Inversions often produce some very intricate patterns. Consider, for example, the following case, illustrated in Figures 4.2.9 through 4.2.12.

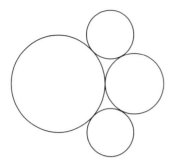

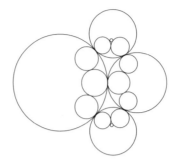

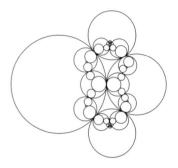

FIGURE 4.2.9 Starting with four touching circles as shown, we apply four inversions with respect to each of them.

FIGURE 4.2.10 Each of these inversions will move the other three circles within the corresponding circles of inversion.

FIGURE 4.2.11 Repeat this procedure, each time inverting with respect to the four starting circles.

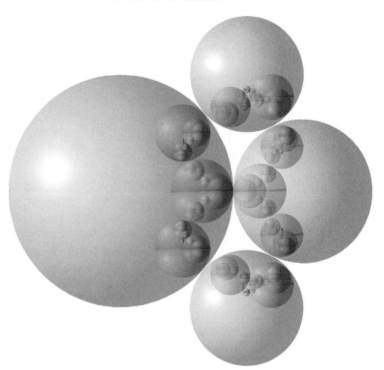

FIGURE 4.2.12 Infinitely many iterations will produce a fractal.

The exploration of the dynamics of inversions (illustrated in Figures 4.2.9 through 4.2.12), as well as the dynamics of linear fractional inversions (mentioned in Section 3.1) and similar (partial) transformations, started by the turn of twentieth century. The esthetic aspects of ϶pt became more prominent and more accessible in the era of computers.

lea of iterating circular inversion (sometimes called circle reflections) infinitely

ιes is an ancient one indeed. *Avatamsaka Sutra* of classical Buddhism describes a network of pearls placed in the heavens by the (Hindu) god Indra:

In the heaven of Indra hangs a net of pearls, made with such skill that each pearl holds the image of every other reflected in its surface, and if you look you see in every shining image again the image of all the others: thus deeper and deeper it goes.

The beautiful book *Indra's Pearls* by David Mumford, Caroline Series, and David Wright explores this subject further. (However, knowledge of the basic theory of complex numbers is a prerequisite for that book.) Figure 4.2.13 is by Jos Leys, an artist who studied the visual potential of the concept of "Indra's pearls."

FIGURE 4.2.13 **(See color insert following page 144.)** Jos Leys, *Indra 461*, 2005.

Exercises:

1. Construct the images of the points *A*, *B*, and *C* under the inversion with respect to the circle H (see Figure 4.2.14).
2. Construct the image of the line *l* under the inversion with respect to the circle H (see Figure 4.2.15). You may use the only theorem in this section, and anticipate that the image of the line is a circle (with one missing point).
3. a. Construct the image of the circle C under the inversion with respect to the circle H (see Figure 4.2.16; what shape will you get? Use the theorem).
   b. Is the image of the center of the circle C under the inversion the center of the image circle *inv*(C)?
4. Construct the image of the circle D under the inversion with respect to the circle H (see Figure 4.2.17; the circles H and D are concentric).

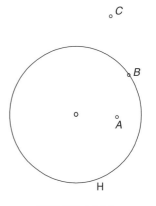

FIGURE 4.2.14

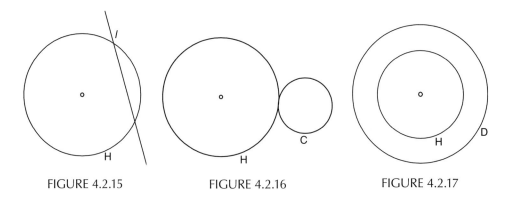

FIGURE 4.2.15   FIGURE 4.2.16   FIGURE 4.2.17

5. Construct the images of the hexagon A and star B under the inversion with respect to the circle H (see Figure 4.2.18).

6. Suppose the inversion circle H is of radius 1. Show that if A is *a* (>0) units away from the center of H, then *inv*(A) is 1/a units away from the center of H. [*Hint*: see the illustration accompanying Step 1 in Inversion 1, and identify two similar triangles.]

7. (*Optional; requires high-school analytic geometry and some patience.*) The goal of this problem is to justify the construction outlined in Figures 4.2.19 through 4.2.22: it is an alternative method for constructing the image of point *A* under the inversion with respect to the circle H.

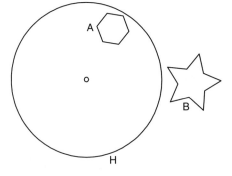

FIGURE 4.2.18

a. Put the starting circle and the point *A* in a coordinate system so that H is centered at the origin and of radius 1 (adjusting the size if necessary), and so that *A* is on the *x*-axis. Assume that *A* has coordinates (a, 0). What are the coordinates of *inv*(A)? [*Hint:* use Exercise 6.]

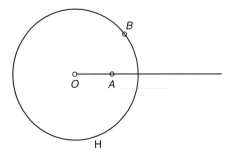

FIGURE 4.2.19 We connect the given point *A* with the center of H and we choose any point *B* so that the angle *AOB* is between −90° and 90°.

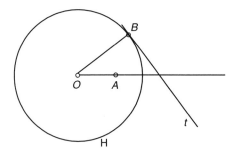

FIGURE 4.2.20 We construct the line through *B* and tangent to H (or, which amounts to the same, the line perpendicular to *OB*).

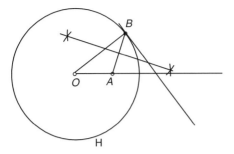

FIGURE 4.2.21 Bisect the segment AB.

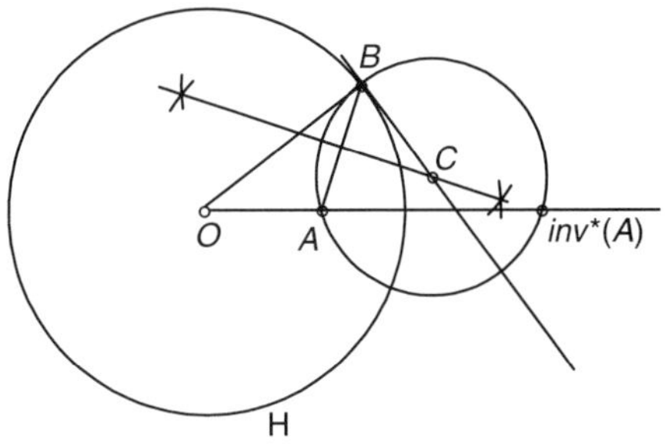

FIGURE 4.2.22 Identify the intersection C of the tangent constructed in the second step and the bisector constructed in the third step. Then draw the circle centered at C and passing through A. The point inv*(A) is then the other intersection point of that circle and the line OA. We indicate in parts (b)–(g) of Exercise 7 how to show that inv(A) = inv*(A).

b. (Refer to Figures 4.2.19 to 4.2.22) Convince yourself that in order to show that inv(A) = inv*(A), it suffices to show that C (Figure 4.2.22) is on the bisector of A and inv(A); that is, see that if C is on the bisector of A and inv(A), then inv(A) and inv*(A) must coincide. Then use Exercise 6 to realize that it suffices to show that the first coordinate of C is $\dfrac{a + (1/a)}{2}$.

c. Suppose B has coordinates (c, d). Given that B is on the unit circle H, what equations do c and d satisfy?

d. Find the equation of the line OB in terms of the coordinates of B. Then use the fact that the tangent line t constructed in Figure 4.2.20 is perpendicular to OB to find the equation of t.

e. Find the coordinates of the midpoint M of AB.

f. Find the equation of the line AB in terms of the coordinates of A and B. Then use the fact that the bisecting line b of the segment AB is perpendicular to AB, to find its equation.

g. Find the first coordinate of the intersection point of the lines t and b (by solving the system of two equations defining t and b) and verify that it is indeed $\dfrac{a + (1/a)}{2}$.

4.3 HYPERBOLIC GEOMETRY

We have learned from the example with the crystal ball that lines could (or, could appear to) be curved (to us), and that for some spaces the usual Euclidean distance does not apply. With that in mind we should not find it hard to accept the following model of hyperbolic geometry, called the **Poincaré model**. (Henri Poincaré was one of the most prominent mathematicians of the early twentieth century.)

The space (the new world) consists of all of the points in the interior of a circle. The circle itself is not a part of the space and it plays the role of a horizon, similar to the role of the contour circle in the crystal ball space viewed from straight below it (shown in

Figure 4.1.5). We will consistently denote that circle by H (for horizon). So, unlike the usual plane, this space can be depicted in whole (Figure 4.3.1).

In order to start dealing with any geometry whatsoever, we need to identify the lines in this space. As you have been warned several times by now, the lines in this space will not be the usual lines. In Figure 4.3.2, we depict four *hyperbolic* lines; one of them is a straight line segment, and the other three are circular arcs. The **first type of hyperbolic lines** in this model consists of all *open* line segments along the diameters of the circle ("open" means that the end points are excluded from the line segments). There is one such hyperbolic line in Figure 4.3.2 (the horizontal line segment depicted in full stroke). The **second type of hyperbolic lines** consists of open circular arcs (again, *open* means that the end points of the arcs within H are excluded) within

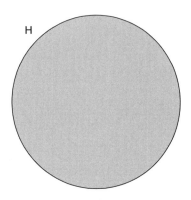

FIGURE 4.3.1 The Poincaré model of the hyperbolic plane geometry consists of the points in the interior of the circle H.

the horizon circle. We take only the arcs that are parts of circles that intersect the horizon H at a right angle (details will be explained soon). In Figure 4.3.2, these hyperbolic lines (in the shapes of circular arcs) are depicted in full stroke, whereas the rest of the circles are in dashed strokes. We will continue applying the terminology we have just introduced: so, *lines* are usual Euclidean lines, whereas *hyperbolic lines* refer to the just described special lines in the Poincaré model of the hyperbolic plane.

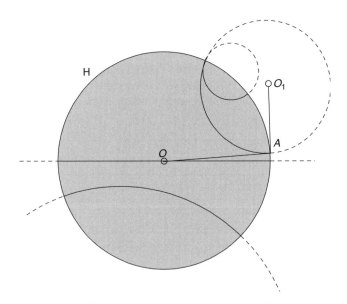

FIGURE 4.3.2 Four hyperbolic lines in the Poincaré model of the hyperbolic plane geometry. In one of them we emphasize one of the features of the circular hyperbolic lines: the line segments OA and AO_1 are perpendicular.

We need to explain the stipulation that the circles we consider should be perpendicular to the horizon circle H. In general, we define the angle between two intersecting smooth curves to be the not larger of the two angles between their tangent lines at the intersection point. A tangent to a smooth curve at a point of the curve is a line that touches the curve at that point (a precise definition requires basic calculus). If the two curves are circles (which is what happens in our story), then the definition simplifies as follows (refer to Figure 4.3.2): the angle between two circles intersecting at a point A is the same as the angle between the two radial segments starting at the centers O and O_1 of the circles and ending at the point A. In particular (and this is something we should keep in mind), two circles are perpendicular (i.e., they intersect at a right angle) if the radial segments ending at the intersection points for the two circles are perpendicular. For example, in Figure 4.3.2 the two radial segments OA and O_1A are perpendicular, and so are the two corresponding circles intersecting at A.

Recall that we are trying to set up a space where the first four of Euclid's postulates are fulfilled, but the fifth one fails, and instead, the fifth hyperbolic postulate is satisfied. At this stage one could prove—we will do that in the next section—that, for example, the first Euclidean postulate holds in this space. That is, it could be shown that given any two points in the hyperbolic space (confined within the circular horizon) there is exactly one *hyperbolic* line passing through these two points. However, we have not yet established everything necessary for us to associate clear meaning to all of the postulates. For example, the second postulate (in Section 1.1) talks about segments of equal *length*; the mention of *length* should alert us, for we realize from the example of the crystal ball world, that *length* in the Poincaré model may not be the same as the usual notion of length. Indeed, that is the case. We describe in the next paragraph a ***hyperbolic length*** in the Poincaré model of geometry (to be used sporadically in the rest of this chapter).

A Bit of Math. Hyperbolic Distance

Take any two points (A and B) in the Poincaré model of the hyperbolic space. As we have already stated, there is exactly one hyperbolic line l that passes through these two points; we keep in mind that hyperbolic lines (look to us as if they) are circular arcs of line segments. Let P and Q be the two intersection points of l (regarded as a usual line or usual circle) with the horizon circle H. Denote the usual distance between any two points X and Y by $d(X, Y)$, and denote the hyperbolic distance we are about to define (when X and Y are in the Poincaré space) by $d_H(X, Y)$. Then $d_H(A, B) = \left| \ln \dfrac{d(A, P) \cdot d(B, Q)}{d(A, Q) \cdot d(B, P)} \right|$ (where ln stands for the natural logarithm—the logarithm with base $e \approx 2.7$). It is not very hard to check that the closer we are to the horizon the larger are the hyperbolic distances between pairs of points of equal usual distance. In other words, if you take two points, say 1 mm (usual) apart from each other and close to the center of a large Poincaré model, then the hyperbolic distance between these two points defined here should not differ much from 1 mm. However, if we move these two points closer to the horizon without altering their usual distance of 1 mm, then the hyperbolic distance increases and could be, say, 1 million km, even though the distance that we *see* has not changed. □

We have almost all of the ingredients needed to be able to deduce that the first four Euclidean postulates are fulfilled. Since that would take us too deep into the realms of axiomatic geometry, we will conveniently choose not to prove some of them. We do pay attention to the fifth postulate. First, we see that the fifth Euclidean postulate fails in this model. In order to establish that, it suffices to find a hyperbolic line and a point out of that line such that it is not true that there is exactly one line through the given point and parallel to the given line. Recall that two (hyperbolic or usual) lines in one plane are parallel if they do not have any common points.

Now, take a look at Figure 4.3.3, and pay attention to the hyperbolic line *C*. The points *O* and *A* are both out of *C*. You can see that the two shown hyperbolic lines (diameters) through *O* are parallel to *C* (since they have no common points with *C*). That is good enough to justify our claim that Euclid's fifth postulate fails. We can also see the same

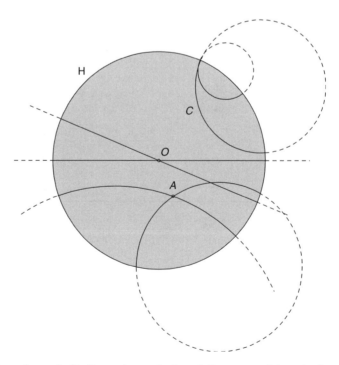

FIGURE 4.3.3 Many hyperbolic lines through *A* and *O* not touching the hyperbolic line *C*.

from the two shown hyperbolic lines (arcs) passing through *A*. All of these hyperbolic lines are parallel to *C*. In fact, one can say that they are even more than parallel to *C* since they do not meet *C* even at the horizon. Such parallel lines are sometimes called hyper-parallel lines (we will use the term *parallel*). The small arc within *C* (Figure 4.3.3) represents a hyperbolic line that is parallel to *C* and that meets *C* at the horizon.

It is visible from Figure 4.3.3 that the fifth *hyperbolic* postulate holds: given any hyperbolic line and any point out of that line, there are infinitely many hyperbolic lines passing through the given point and parallel to the given line. For example, it is clear that there are infinitely many hyperbolic lines (diameters) through the center *O* that are parallel to (not touching) the hyperbolic line *C*. Once we describe a procedure for constructing general hyperbolic lines through any given point in the hyperbolic plane, we will see how to get infinitely many hyperbolic lines parallel to a given hyperbolic line, and passing through a point other than the center of the horizon circle.

Not surprisingly, some other properties (aside from the postulate on parallel lines) in the classical, Euclidean geometry also fail in the setting of the hyperbolic geometry. For example, it follows from Euclid's five postulates that the sum of the interior angles of any triangle is 180°. In fact, this particular property of the angles of a triangle is *equivalent* to Euclid's fifth postulate, which means that not only can we deduce it from Euclid's fifth postulate (and the other four postulates) but we can also deduce the fifth postulate from that property. Consequently, that particular property *must* fail in the hyperbolic geometry, for if it is true, then both the fifth Euclidean postulate and its negation would be true, a clear contradiction. So, in hyperbolic geometry there are triangles with the sum of interior angles not equal to 180°. In fact, more is true: the sum of the interior angles in *any* triangle in the Poincaré model of hyperbolic geometry must be strictly less than 180°.

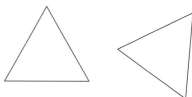

In Figure 4.3.4, we see two triangles. The sum of the interior angles in each of the triangles is 180°. We have no problem accepting that the two triangles are congruent (of the same size). In Figure 4.3.5, we also see two triangles. You can measure the interior angles if you wish—the sum of these angles in both cases is less than 180°. Each of the two triangles is an equilateral triangle: the three sides are of equal hyperbolic length. What may surprise you (if you

FIGURE 4.3.4 Two congruent equilateral triangles.

rely mostly on your visual perception) is that the two triangles to the right are also congruent (of the same size). Do not forget the hyperbolic distance is different from the usual, Euclidean distance.

This property (that the sum of the angles in a triangle is less than 180°) is precisely why hyperbolic geometry is interesting from the visual point of view. For example, we will see in the last section of this chapter that one consequence of that property is that we have *infinitely* many *regular monohedral* tilings (defined in Section 2.5) of the hyperbolic plane, as opposed to only three such tilings in the Euclidean plane.

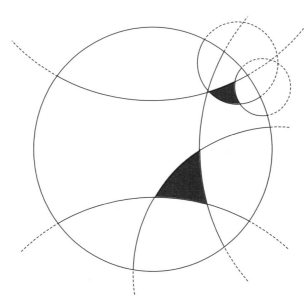

FIGURE 4.3.5 Two congruent equilateral hyperbolic triangles.

Exercises:

1. Use a protractor to measure the interior angles of the hyperbolic triangles in Figure 4.3.5 and find the sum of all interior angles in each of the triangles. (Note: you first need to construct the tangents.)

2. Let O be the center of the circle in the Poincaré model of a hyperbolic planar geometry, and let A be any other point in that circle.
   a. Construct a hyperbolic line passing through O and A.
   b. Show that no circle passing through O determines a hyperbolic line.
   c. Show Euclid's first postulate in the case of these two points. That is, show that there is exactly one hyperbolic line passing through O and A.

3. Suppose the horizon in the Poincaré model is a unit circle. Find the hyperbolic distance between the center of the horizon circle and a point 0.5 units away from the center. (You may need a calculator.)

4. Given the horizon circle of a Poincaré model, and given the center O of the horizon circle:
   a. Construct one hyperbolic line that does not pass through O
   b. Construct one hyperbolic triangle

5. Construct one hyperbolic circle. [*Hint*: recall that a circle of radius r is the set of points that are at a distance r from a fixed point (the center of the circle). Where should we choose the center of the hyperbolic circle to get an easy construction?]

4.4 SOME BASIC CONSTRUCTIONS IN THE POINCARÉ MODEL

We will exclusively deal with the Poincaré model of hyperbolic geometry, and for the sake of simplicity from now on we will call it the **hyperbolic plane**. The following basic constructions of various objects in the hyperbolic plane will clarify some of the outstanding questions we have stated in the previous sections. As usual, *construct* means *construct using an unmarked ruler and a compass*.

CONSTRUCTION 1. CONSTRUCTING A HYPERBOLIC LINE THROUGH ANY GIVEN POINT; IDENTIFYING ALL HYPERBOLIC LINES PASSING THROUGH A GIVEN POINT

We are given the hyperbolic plane and a point A in the hyperbolic plane (Figure 4.4.1). We know from Section 1.2 how to construct the center O of the horizon circle. So, we can assume that Figure 4.4.1 represents the initial data. It is easy to construct one hyperbolic line: simply join O and A and extend the segment to the (open) diameter of the circle H. Our goal is more ambitious: we want to see how we can construct *every* hyperbolic line passing through A.

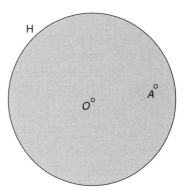

FIGURE 4.4.1

The construction of all the hyperbolic lines through A is closely related to inversion with respect to the circle H. Indeed, the first, and the largest step in the construction of hyperbolic lines through A is the construction of the image $inv(A)$ of A under the inversion with respect to H. (See Section 4.2; the construction outlined in Figure 4.4.2 here.) Once we have $inv(A)$ we simply find the bisector c_A of the line segment from A to $inv(A)$.

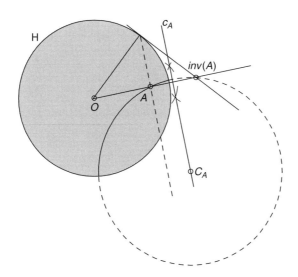

FIGURE 4.4.2 The bisector c_A of A and $inv(A)$ contains all the centers of the hyperbolic lines through A.

Here is the final and main aspect of the construction: *every* circle centered at the line c_A and passing through the point A intersects the horizon circle H at a right angle, and, conversely, every circle passing through A and intersecting H at a right angle must have its center at the line c_A. (We will not prove these two claims; we only note that a part of the justification is hidden within the solution of the longish Exercise 7, Section 4.2.) Consequently, every such circle gives rise to a hyperbolic line passing through the point A. In Figure 4.4.2 you can see one such circle, centered at the point C_A on the line c_A. (It passes through A, and so passes through $inv(A)$ as well.) ☐

Now we can explain why we claimed earlier that the hyperbolic lines that passed through the center O of the horizon circle could be considered a special case of the general *curved* hyperbolic lines, even though the former were line segments while the latter were circular arcs. Suppose we choose points on c_A farther and farther from A. Then the circles centered at these points and passing through A are larger and larger. In case of such an extremely large circle, the hyperbolic line (the part of that circle within H) will be almost identical to the hyperbolic line defined by the diameter of H passing through O and A. So, we can say that the hyperbolic lines on the diameters of the circle H are in fact circles with infinitely large radii. In particular, the hyperbolic line we observed first (the diameter through O and A), is, loosely speaking, a part of a circle with center at the line c_A and with infinite radius.

In order to shorten the subsequent explanations, the line c_A containing all the centers of the hyperbolic lines through A will be tentatively called the ***centerline*** for A.

CONSTRUCTION 2. CONSTRUCTING THE UNIQUE HYPERBOLIC LINE THROUGH TWO GIVEN POINTS

We are given two points in the hyperbolic plane and we need to construct a hyperbolic line passing through both the points. It will be clear from the construction that there is exactly one such line, thus proving that the first postulate of the Euclidean and the hyperbolic geometry holds in this (Poincaré) model.

The starting setting consists of the hyperbolic plane and the points A and B (refer to Figure 4.4.3). We construct the centerline for the point A (line c_A in Figure 4.4.3) and the

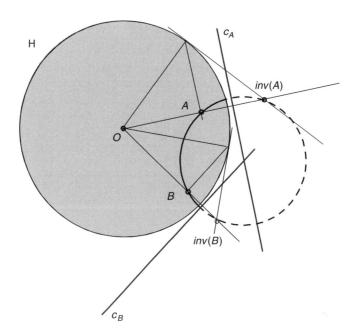

FIGURE 4.4.3 The center of the hyperbolic line passing through A and B is at the intersection of the lines c_B and c_A.

centerline for the point B (line c_B) by following the procedure in Construction 1. Every circle giving rise to a hyperbolic line through A has its center on c_A, and every circle giving rise to a hyperbolic line through B has its center on c_B. It follows that the center of a circle determining a hyperbolic line through both A and B must be in the intersection point of c_A and c_B. If the points O, A, and B are not on the same line, then the lines c_A and c_B have exactly one intersection point—so that we have uniqueness (i.e., we can talk about **the** center of the circle passing through A and B). The case when O, A, and B are on the same line is left as an exercise. So, finishing off our construction, we now need to draw a circle centered at the intersection point of c_A and c_B and passing through both A and B (Figure 4.4.3). $\square$

Note that there is a somewhat shorter route to the center of the hyperbolic line through A and B: once we get c_A, we need to construct only the bisecting line b of the line segment AB. The center is then at the intersection of the lines c_A and b.

CONSTRUCTION 3. CONSTRUCTING A HYPERBOLIC LINE INTERSECTING ANOTHER HYPERBOLIC LINE AT A GIVEN ANGLE

We are given an angle (the angle called α in Figure 4.4.4), a hyperbolic line (denoted l), and a point A on l. We need to construct a hyperbolic line passing through A and intersecting l at the given angle α. For the sake of readability, we subdivide the construction into smaller steps.

Step 1. Draw a tangent to l at the point A. Since l is a part of a circle, we know how to find its center (Section 1.2). As we noted in Section 4.2 (see Step 3 in Inversion 1), the tangent line at A is simply the line passing through A and perpendicular to the line segment from the center of the circle supporting l to the point A. Denote this tangent line by t (illustrated in Step 1 below).

Step 2. Draw one line (denoted m) passing through A and intersecting t at the angle α. There are two such lines and so there will eventually be two solutions to this prob-

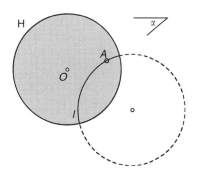

FIGURE 4.4.4 The setup: we are given a hyperbolic line l and an angle α.

lem. We only show one of them. Recall again that drawing m amounts to replicating the angle α at the point A, with one side of the angle being t (Section 1.2).

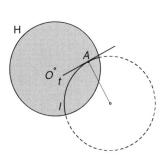

STEP 1. Constructing the tangent line t.

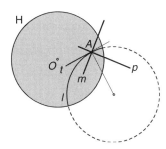

STEPS 2 and 3.

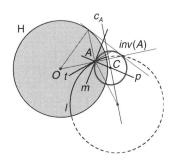

STEPS 4 and 5.

Step 3. Draw the perpendicular to *m* at *A*. The center of the circle supporting the hyperbolic line we need to construct is somewhere on that perpendicular line. We denote that perpendicular by *p*.

Step 4. Find the centerline c_A for *A*. We know that every circle supporting the hyperbolic line through *A* must be on that centerline.

Step 5. Find the intersection point *C* of the centerline c_A for *A* and the line *p*. This is the center of the hyperbolic line we are constructing. Now simply draw the circle centered at *C* and passing through *A*. The part of that circle within the horizon line *H* is one of the two hyperbolic lines intersecting the given hyperbolic line *l* at *A* and at the angle *α*. □

CONSTRUCTION 4. CONSTRUCTING ONE EQUILATERAL TRIANGLE

The following simple construction of an equilateral hyperbolic triangle is of some importance to us as a preview to the next section.

We already saw in the previous section that equilateral triangles in the hyperbolic plane may appear to have sides of different length, and we know that this happens because *hyperbolic lengths* are not the same as the usual Euclidean lengths. Because of that, various simple constructions in the Euclidean plane become much more complicated in the hyperbolic plane. For example, finding the midpoint of a hyperbolic line segment must take into account the hyperbolic length and thereby is not as simple as finding the midpoints of usual line segments.

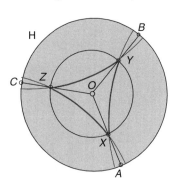

FIGURE 4.4.5 An equilateral hyperbolic triangle *XYZ*.

In order to avoid extensive reliance on the complicated formula for the hyperbolic length that we have introduced in Section 4.3, we base our construction and its justification on *symmetry*.

Step 1. Subdivide the horizon circle *H* into three congruent arcs. This is same as subdividing an angle of 360° into three equal parts (so each of these three angles is 120°). We know how to construct an angle of 60° and constructing 120° is just doubling a 60° angle (Exercises in Section 1.2). In Figure 4.4.5, the angles ∠AOB, ∠BOC, and ∠COA are all 120°.

(Side problem: can you construct an angle of 120° directly, without first constructing 60°?)

Step 2. Join the center *O* with the three points *A*, *B*, and *C* to get three line segments. Choose a point *X* between *O* and *A* and draw a circle with radius *OX* and centered at *O*. This circle intersects the other two segments *OB* and *OC* at the points *Y* and *Z*.

Step 3. Construct the hyperbolic line through each pair of the three points *X*, *Y*, and *Z*. The details are given in Construction 2 in this section. The hyperbolic line segments from *X* to *Y*, from *Y* to *Z*, and from *Z* to *X* are the sides of an equilateral hyperbolic triangle. The construction is thus completed.

Step 4. (*optional; justification*). The hyperbolic triangle we get (with sides along the hyperbolic lines *XY*, *YZ*, and *ZX*) is equilateral. This follows from the symmetry of our picture and since the definition of hyperbolic length given in Section 4.3 respects that symmetry. More to the point, we need to show that $d_H(X, Y) = d_H(Y, Z) = d_H(Z, X)$. According to our

definition of the hyperbolic distance, we have $d_H(X, Y) = \left| \ln \dfrac{d(X, A) \cdot d(Y, B)}{d(X, B) \cdot d(Y, A)} \right|$ and $d_H(Y, Z) =$

$\left| \ln \dfrac{d(Y, B) \cdot d(Z, C)}{d(Y, C) \cdot d(Z, B)} \right|$.

But we see from the picture and it follows from our construction that $d(X, A) = d(Y, B) = d(Z, C)$ and that $d(X, B) = d(Y, C) = d(Y, A) = d(Z, B)$, so that $d_H(X, Y) = d_H(Y, Z)$. Similarly $d_H(Y, Z) = d_H(Z, X)$, and so all the sides of the hyperbolic triangle are of equal hyperbolic length, as claimed. □

Exercises:

1. Construct two hyperbolic lines passing through the center of the horizon circle and making an angle of 45°.

2. Construct one hyperbolic circle. (A hyperbolic circle of radius r and centered at a point P is the set of all points in the hyperbolic plane at hyperbolic distance r from P.)

3. We are given two points A and B in the hyperbolic plane such that the center O of the horizon circle and the points A and B are all distinct points lying on the same line. Construct the hyperbolic line passing through A and B. Why is this hyperbolic line unique?

4. We are given a point A in the hyperbolic plane (Figure 4.4.6). Construct two hyperbolic lines passing through A and intersecting each other at an angle of 60°.

5. a. Construct one hyperbolic square.
   b. Construct one hyperbolic regular pentagon.

6. Given two points A and B in the hyperbolic plane, construct a pair of parallel lines, one passing through A, the other through B, such that these two lines meet at the horizon.

7. Construct the hyperbolic midpoint of the hyperbolic line segment XY in Figure 4.4.5. Justify your answer by using the formula for hyperbolic distance.

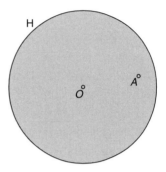

FIGURE 4.4.6

4.5 TILINGS OF THE HYPERBOLIC PLANE

Recall that in the Poincaré model of the hyperbolic geometry the sum of the angles in a triangle is always strictly less than 180°. Moreover, given any angle α between 0° and 180°, there is a hyperbolic triangle with the sum of the interior angles equal to α. For example, there is a hyperbolic triangle with the sum of the interior angles equal to 1°. This is relatively easy to justify, and we do it in the next few sentences and Figure 4.5.1.

Recall the construction of an equilateral triangle done in the previous section and consider what happens when the circle S through the points X, Y, and Z is chosen to be closer and closer to the horizon (see Figure 4.5.1), so that X, Y, and Z approach positions A, B, and C, respectively. As S gets closer and closer to the horizon, the sides of the triangle

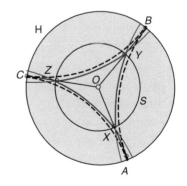

FIGURE 4.5.1 As we increase the radius of the circle S, the points X, Y, and Z approach the horizon, and the interior angles of the hyperbolic triangle XYZ become smaller and smaller, approaching 0°.

get closer and closer to intersect at the points *A*, *B*, and *C* and the interior angles of the hyperbolic triangle decrease, each of them becoming closer and closer to 0°. The last claim follows from the fact that the dashed arcs in Figure 4.5.1 intersect the horizon at right angles, and thereby are mutually tangential (touch each other) at these intersection points. So the intersection (touching) angles for the dashed circles are all 0°.

The equilateral triangles with vertices at the points *X*, *Y*, and *Z* (as *X*, *Y*, and *Z* approach the horizon) are symmetric with respect to the center of the Poincaré disk. The reason we have chosen triangles positioned in such a way is purely practical—we wanted to utilize the symmetry of the hyperbolic plane rather than rely on the formula for hyperbolic distances—but the reader should not assume that what we perceive as the center of the hyperbolic plane has some special role within the hyperbolic plane. Indeed, for a two-dimensional creature living in this world, the center *O* has no special meaning at all and is nothing but one of the infinitely many points, no more, no less important—any other point is also of infinite *hyperbolic* distance from the horizon circle H.

In Euclidean geometry, each interior angle of an equilateral triangle is exactly 60° and so we can place at most six nonoverlapping equilateral triangles around a vertex in the (usual) plane. Similarly, we can have at most four nonoverlapping squares around a vertex, and at most three nonoverlapping regular hexagons around a vertex. The fixed sizes of the interior angles of the regular polygons in the Euclidean geometry preclude using regular polygons with many sides as tiles in regular tilings. For example, we cannot use any regular polygon with, say, 13 sides, since the peculiar size of the interior angle of such a polygon (about 76.15°) would not allow us to cover the neighborhood of a vertex with such nonoverlapping regular polygons. These are the main reasons why we have only a few regular tilings, and why we do not have many more semiregular tilings of the (usual) plane. However, in hyperbolic geometry there are no such constraints. For example, since equilateral triangles could have as small interior angles as we please, we can place any number of nonoverlapping congruent hyperbolic equilateral triangles around a point in the hyperbolic plane. Similarly, we can place as many nonoverlapping regular hyperbolic polygons as we please (with sides of equal hyperbolic distance and with equal interior angles) around any point. As a consequence, the hyperbolic geometry is much richer with tilings of various types.

We start with a few pictures (Figures 4.5.2 through 4.5.5) of various tilings of the hyperbolic plane and then we will see (Figures 4.5.6 through 4.5.10) how these or similar tilings have been used in art.

The tilings of the hyperbolic plane and other visual aspects of hyperbolic geometry were unknown to Escher until he met the geometer H. S. M. Coxeter. Escher was inspired by an illustration in a Coxeter's book depicting a tiling of the hyperbolic plane and used the idea to produce a series of woodcuts and prints entitled *Circle Limit 1, 2, 3 and 4*. The technique Escher uses to get the underlying drawings is in essence the same as the one described in Section 2.5.

The connection between the images shown in Figures 4.5.8 and 4.5.9 is emphasized in Figure 4.5.10, where we superimpose the skeleton of the image shown in Figure 4.5.8 (slightly rotated) over Escher's artwork in Figure 4.5.9. In Figure 4.5.11, we show an artwork by Jos Leys based on a 4,4,4,4,4,4-tiling of the hyperbolic plane.

FIGURE 4.5.2 Hyperbolic tiling 4, 4, 4, 6. The notation indicates that we encounter three squares and one regular hexagon around every vertex. Note that in this setting a *square* is a regular polygon with four sides of equal hyperbolic length and with four equal interior angles, none of them 90°.

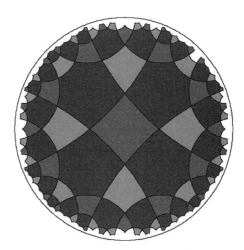

FIGURE 4.5.3 Hyperbolic tiling 4, 6, 4, 6: this time there is a square, a regular hexagon, again a square, and one more regular hexagon around every vertex. These types of tilings do not have a counterpart in the Euclidean geometry.

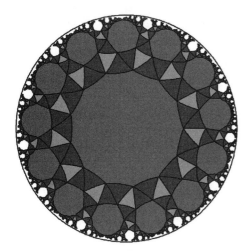

FIGURE 4.5.4 Hyperbolic tiling 13, 3, 3, 3, 3. Unlike Euclidean geometry there are no obstacles in using a regular polygon with 13 sides in a tiling. We use it above, in conjunction with equilateral triangles.

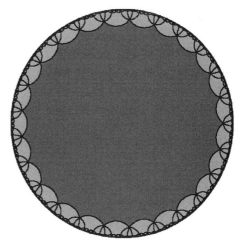

FIGURE 4.5.5 Hyperbolic tiling 25, 5, 5, 5, 5, 5. A tiling with a regular polygon with 25 many sides, combined with regular pentagons. Not much can be seen close to the horizon (i.e., far away in the hyperbolic sense).

FIGURE 4.5.6 Hyperbolic tiling 4, 3, 4, 3, 4, 3. The reason we display this tiling is the picture to the right.

FIGURE 4.5.7 M. C. Escher. *Circle limit 3*, 1959, woodcut. The underlying tiling is visible and we see that it is the same as the one depicted to the left.

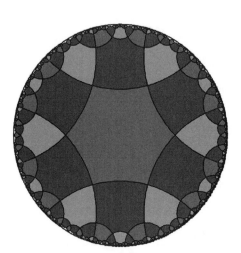

FIGURE 4.5.8 A tiling 6, 6, 6, 6 of the hyperbolic plane.

FIGURE 4.5.9 M. C. Escher. *Circle limit 4*, woodcut, 1960. (This woodcut is known as *Heaven and Hell*, and also known as *Angels and Devils*.)

FIGURE 4.5.10

FIGURE 4.5.11 **(See color insert following page 144.)** Jos Leys, *hyp 204*, 2002.

CHAPTER 5

Perspective and Some Three-Dimensional Objects

We begin this chapter by discussing the basic rules of perspective (Section 5.1), and then we briefly analyze these rules from the mathematical point of view (Section 5.2). In Sections 5.3 and 5.4, we consider various types of three-dimensional objects, including polyhedra, spheres, cylinders, and cones. In Section 5.4, we classify curves that are intersections of cones with planes (the conic sections) and introduce methods for sketching such curves. In the last section, we present a brief overview of some three-dimensional analogs of objects and notions we have covered earlier.

5.1 PERSPECTIVE

Perspective drawing was a technique developed during the Renaissance to produce a true, "visual" geometry. The main objective was to draw three-dimensional objects faithfully onto a two-dimensional canvas so that we could recover the three-dimensional aspects of the original scene. As humans, we are, in some way, built to perceive the three-dimensionality of the surrounding world—indeed we presuppose that we are three-dimensional beings. However, our basic visual perception is inherently two-dimensional. *Seeing* is a projection of the three-dimensional images around us onto the eye retina, with the rest of the process of visualizing the three dimensions done by our brain (analyzing the images and associating distances, order, sizes, shadows, and other properties of the objects we see). We recognize the three-dimensionality by *decoding* the two-dimensional images using knowledge that we mostly acquire in our childhood. Indeed, some studies have shown that people who have gained sight later in their life have trouble in interpreting the 2D-to-3D code and reconstructing distances from it.

Drawing (two-dimensional) realistic pictures of three-dimensional objects is equivalent to projecting the images of these three-dimensional objects (that we see or imagine)

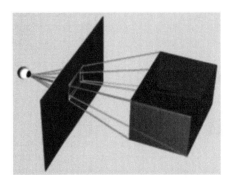

FIGURE 5.1.1 and Animation 5.1.1.

on the picture plane. The corresponding process of projecting images of objects is called ***perspective drawing*** (perspective painting, or similar) and it is illustrated in the following graphics (Figure 5.1.1 and Animation 5.1.1). The red lines in the graphics show some rays connecting the eye with some strategically chosen points in the 3-D scene that we are depicting. The intersections of these rays with the drawing plane are the perspective images of the points on the drawing plane. A renaissance technique for perspective drawing is shown in Figure 5.1.2.

FIGURE 5.1.2 In this illustration by Albrecht Dürer (1525), we see an artist making a sketch of the scene. As in Figure 5.1.1/Animation 5.1.1, the drawing plane is transparent (made of glass), facilitating the location of the perspective images of the main points of the scene.

Basic Rules of Perspective Representation

The principles of (classical) perspective drawing can be summarized in the following list of four basic rules of *classical* perspective. (There exists a nonclassical, *curvilinear* perspective; we will not cover this topic.) In three of the four rules, we will discuss mutually parallel lines in three dimensions. Two lines in three dimensions are parallel if they do not intersect and if they lie on a single plane. Lines that do not intersect and are not parallel are called skew lines.

Rule 1. The farther an object is from the drawing plane, the smaller is its perspective image in the drawing plane (Figures 5.1.3 and 5.1.4).

FIGURE 5.1.3 Here is a row of copies of Dadat. When we decode this picture and reconstruct the three-dimensional scene, all copies of Dadat seem to be of the same size; ....

FIGURE 5.1.4 ... but *in* the picture they are not; see how the last Dadat compares with the first one, without altering their sizes.

It may seem that Rule 1 is too obvious—however, as we will see, it is an essential rule that we often take for granted. Precisely, because of this rule it is easy to produce visual paradoxes by consciously and cleverly using it or violating it. Such examples appear by the end of this section (Figures 5.1.25 and 5.1.26).

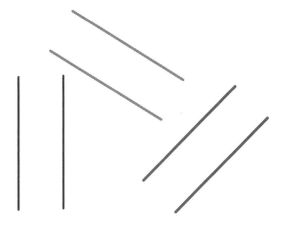

FIGURE 5.1.5 These lines are parallel to the picture plane. In such cases, each pair of parallel lines is depicted as a pair of parallel lines.

Rule 2. Lines that are mutually parallel and parallel to the painting plane are depicted as parallel lines (Figure 5.1.5).

Rule 3. Parallel lines that are not parallel to the painting plane are depicted as lines intersecting at a single point called a ***vanishing point*** (Figures 5.1.6 through 5.1.8).

 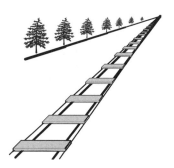 

FIGURE 5.1.6 Three inter-
secting lines representing
parallel lines in three
dimensions.

FIGURE 5.1.7 It is easy to
associate three dimen-
sions if we add a few more
elements to the same three
lines. Now we *see* that they
are *really* parallel.

FIGURE 5.1.8 Train tracks.

We introduce a bit of terminology: a set of two or more parallel lines is called a *class of parallel lines*. Rule 3 then stipulates that each class of parallel lines that is not parallel to the drawing plane determines a vanishing point.

Rule 4. If three or more classes of parallel lines are all parallel to a fixed plane that is not parallel to the drawing plane, and if these classes of parallel lines determine three or more vanishing points, then all of these vanishing points occur on a single line (Figure 5.1.9).

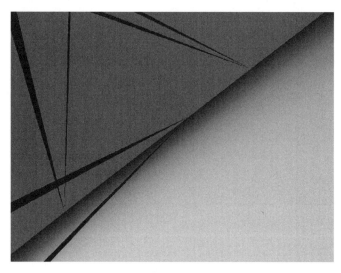

FIGURE 5.1.9 Here are three classes of parallel lines, each class consisting of a pair of lines, and all of the lines are parallel to the yellowish plane at the right bottom. The two lines in each of the pairs are mutually parallel in the three-dimensional scene. They are depicted as intersecting lines, and the three intersection points all lie on a single line.

If in addition to the assumptions in the last rule, all lines are horizontal, then the line containing the three intersecting points is called the ***horizon line*** (Figure 5.1.10).

FIGURE 5.1.10 All lines in this case are parallel to the horizontal yellowish plane at the bottom. Pairs of mutually parallel lines of this type are depicted as intersecting lines, and all of the intersection points lie on the horizon line. □

A Short Note on the History of Perspective Drawing

We digress a bit to take a very brief look at the history of perspective.

The icon shown in Figure 5.1.11 is from the pre-Renaissance period. There are some obvious incongruities related to distances and visibility in the painting, perhaps due to the desire of the author to give prominence to some features of the painting at the expense

FIGURE 5.1.11 **(See color insert following page 144.)** *The Annunciation*, fourteenth century, Ohrid, Macedonia.

of others. For example, notice that the two front pillars of the throne appear both in front and behind the image of Mary. Note also that parallel edges of the throne and around the throne are depicted as parallel lines, even though they are not parallel to the drawing plane (violating Rule 3).

The icon in Figure 5.1.12 is a valiant, preperspective attempt to introduce some perspective elements in the scene. Most notably, parallel edges are depicted as nonparallel (as they should), and, as we see from the small analysis in Figure 5.1.13, some classes of parallel lines are depicted as intersecting at the corresponding vanishing points. However, equidistant pairs of points on the edges of the flat roof of the buildings in the background are depicted in such a way that those that are farther are shown as *larger*, violating Rule 1. As a consequence, the 3-D position of the corresponding vanishing point seems to be closer to the viewer of the scene; instead, it should be farther away.

FIGURE 5.1.12 **(See color insert following page 144.)** *The Assumption of Virgin Mary*, fourteenth century, Ohrid, Macedonia.

FIGURE 5.1.13 There is a vanishing point for the edges of the roof, but it is on the wrong side of the building: it should be on the far side of the building, not in front of it.

The piece of art shown in Figure 5.1.14 is from the early Renaissance period.

The painting shown in Figure 5.1.14 was done about a century after the icons shown in Figures 5.1.11 and 5.1.12 had been completed. We see the progress in the attempt to faithfully depict reality. The author employed *intuitive perspective*: he was aware that some parallel lines should be depicted as *receding* intersecting lines. However, as pointed out in the caption for Figure 5.1.15, some classes of parallel lines are not depicted as intersecting at a single *vanishing* point, as they should.

FIGURE 5.1.14 **(See color insert following page 144)** Jean Fouquet, *The royal banquet,* detail from *Les grandes chroniques de France,* 1420–1480.

FIGURE 5.1.15 The receding lines of the floor tiling are all mutually parallel in the three-dimensional scene. According to Rules 3 and 4 they should be depicted as lines intersecting at a single point (the vanishing point). We see that this does not happen. There are other violations of the rules if we consider the parallel edges of the ceiling.

As a representative of Renaissance painters we consider Leonardo da Vinci. The sketch in Figure 5.1.16 was used as an underpainting for the final version of da Vinci's masterpiece, and shows the painstaking details the Renaissance painters went through in their quest to faithfully reproduce the real world.

The discovery of the principles of the (classical) perspective representation of objects is usually attributed to Filippo Brunelleschi (1377–1446), who was an architect and a goldsmith. Unfortunately, the original two studies where he presented his invention are now lost. □

One-Point, Two-Point, and Three-Point Perspective

Each cube (or parallelopiped) contains three classes of mutually parallel edges. There are three possible positions of the cube with respect to the picture plane, depending on these three classes of parallel edges (lines).

    a. Two of the three classes of parallel edges are parallel to the picture plane.
    b. Only one class of parallel edges is parallel to the picture plane.
    c. None of the edges on the cube are parallel to the picture plane.

FIGURE 5.1.16 Leonardo da Vinci, *The Adoration of the Magi*, sketch, 1582.

These three possibilities determine the type of perspective: (a) one-point perspective, (b) two-point perspective, and (c) three-point perspective. More explanation is given in the captions of Figures 5.1.17 through 5.1.19.

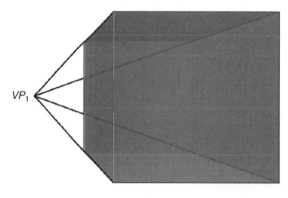

FIGURE 5.1.17 **One-point perspective:** observe that only four of the edges of the cube are not parallel to the painting plane. These four edges belong to a single class of parallel lines, depicted as intersecting at the only vanishing point.

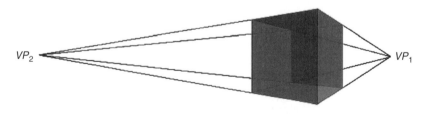

FIGURE 5.1.18 **Two-point perspective:** only the class of *vertical* lines is parallel to the painting plane. The other two classes of parallel lines determine the two vanishing points.

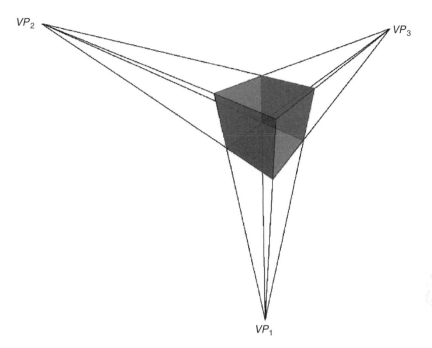

FIGURE 5.1.19 ***Three-point perspective***: none of the three classes of mutually parallel edges is parallel to the painting plane. There are three vanishing points, one per each class of parallel edges.

Basic Example. Finding the Horizon Line

Our goal in this example is to construct the horizon line corresponding to the perspective drawing of a house shown in Figure 5.1.20.

FIGURE 5.1.20

Solution. Mutually parallel horizontal lines should be depicted as intersecting at the horizon in (two-dimensional) pictures. To identify the horizon we need two pairs of mutually parallel horizontal lines having two different vanishing points (i.e., we need two classes of parallel lines). These two pairs are shown in Figure 5.1.21. The two intersections yield two vanishing points and the horizon line passing through them.

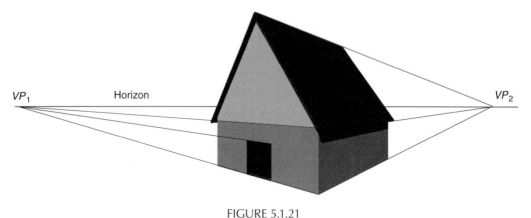

FIGURE 5.1.21

Basic Example. Drawing a Two-Point Perspective Image of a Square

In Figures 5.1.22 through 5.1.24, we describe a simple procedure for drawing a two-point perspective image of a square.

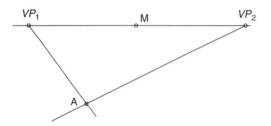

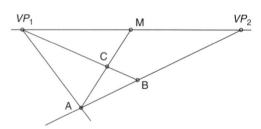

FIGURE 5.1.22 Start with a horizon line, any two vanishing points, and find (construct) the midpoint M of the line segment between the vanishing points. Choose a point, and draw the lines through A and the vanishing points. The point A will be the vertex of the square closest to the drawing plane.

FIGURE 5.1.23 Chose the point B anywhere on the line segment from A to VP_2. Join B and VP_1, and join A and M. The intersection of these two lines is the point denoted by C. The points B and C are further two vertices of the square we draw.

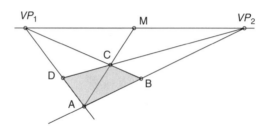

FIGURE 5.1.24 Connect C with the vanishing point VP_2 and identify the intersection point D. Then $ABCD$ is a square drawn in two-point perspective. □

Impossible Worlds—Diversions in Perspective Drawings

Here are two masters playing around with the rules of perspective (see Figures 5.1.25 and 5.1.26). Are some of the rules of perspective violated? If yes, which ones and how?

FIGURE 5.1.25 Jos de May, *Carefully restored Roman ruin in a forgotten Flemish locality with Oriental influences*, china ink drawing, 1983 © Jos de Mey, Zomergem, Belgium.

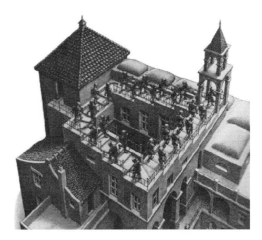

FIGURE 5.1.26 M. C. Escher. *Ascending and descending*, lithograph, 1960. The idea comes from an article by L. S. Penrose and R. Penrose (1956, father and son); we referred to R. Penrose when we discussed aperiodic tilings in Section 2.4.

Surprisingly, there is nothing wrong with either of the pictures regarding Rules 2, 3, and 4. Even more surprisingly, under certain assumptions (to be clarified below) not even Rule 1 is violated. So then, what has happened? As we will explain below, it is true that Rule 1 is the main cause behind the obvious difficulties we encounter when we have to reconstruct the three-dimensional scenes from the above two-dimensional representations—but the problem lies more with us and with the implicit (subconscious) algorithms we employ to decode three-dimensionality from two dimensions.

In Figure 5.1.27, we show a simulation of Escher's (Penrose's) idea; we analyze it in Figures 5.1.28 through 5.1.30.

In Figure 5.1.29, we see clear vertical discrepancies between copies of Dadat: some are far above the surface of the water, whereas others are close to the surface of the ocean. The latter are much larger in the three-dimensional scene. If we assume that the Dadat at the topmost stairs is 182 cm (6 ft.) then the Dadat in the water is a giant of 273 cm (9 ft.)—they only appear to be of the same size in Figure 5.1.27. So, when we decode the three-dimensional scene from this picture of "ever-ascending" little people, we have a choice to make: either we assume that the 11 Dadats are of the same size (in which case

FIGURE 5.1.27 There are 11 copies of Dadat and they are all climbing up along a circular stairway. Is this a visual paradox? Not really—see the next pictures.

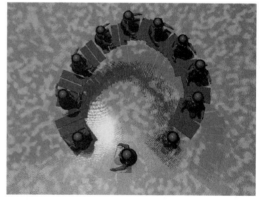

FIGURE 5.1.28 Setting the same scene over water we see something strange happening with the shadows.

FIGURE 5.1.29 The mystery is solved the moment we raise the level of water.

we do get an eternal ascending "paradox"), or we accept that they are not of the same size. It seems that our decoding algorithm is so overwhelmed by our *experience* that it limits our choices and virtually forces us to adopt the first interpretation almost without exceptions. One can claim that most of the so-called visual paradoxes are based on the above trick. □

FIGURE 5.1.30 If we change the viewpoint we clearly see what has happened: the lowermost Dadat and the topmost Dadat are of different sizes, but when we choose a high viewpoint so that they are aligned with the viewpoint (as we did in the above pictures) they appear to be of the same size.

Exercises:

1. a. Consider the icon in Figure 5.1.11: explain how the rules of perspective have been violated.
   b. Do the same as in (a) for Fouquet's picture (Figure 5.1.14).
2. Complete the one-point perspective image given in Figure 5.1.31 by drawing windows on the sidewalls, as well as pieces of furniture of your choice.
3. The topside and one of the (front) vertical edges of a two-point perspective drawing of a box are given in Figure 5.1.32. Copy the illustration and then find the vanishing points and draw the rest of the box.

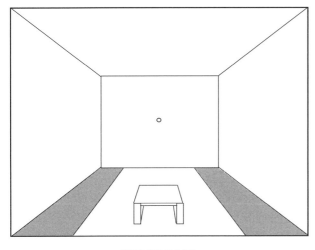

FIGURE 5.1.31

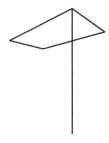

FIGURE 5.1.32

4. Figure 5.1.33 shows a vanishing point, the horizon, and three edges of a box in perspective drawing (the three edges that are shown correspond to the thick edges in the following

cube ). Identify the second vanishing point and draw the rest of the box.

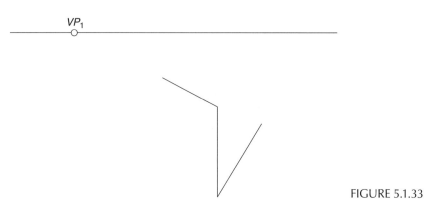

FIGURE 5.1.33

5. In Figure 5.1.34, four vertices of a box are drawn in a three-point perspective. (In the small cube shown at the bottom left of Figure 5.1.34, we indicate which four vertices are drawn.) The three vanishing points VP_1, VP_2, and VP_3 are also given. Draw the rest of the box in three-point perspective.

6*. The picture in Figure 5.1.35 depicts the outer edges of a chessboard in a one-point perspective; as you know, a chessboard is made of $8 \times 8 = 64$ squares. Draw the rest of the chessboard. [*Hint*: start with two diagonals.]

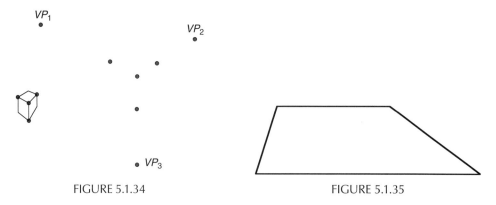

FIGURE 5.1.34   FIGURE 5.1.35

7. The graphic in Figure 5.1.36 gives the perspective image of a box with a cover. Draw the vanishing points corresponding to all pairs of parallel lines passing through various edges of the box.

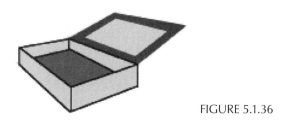

FIGURE 5.1.36

8. Figure 5.1.37 shows a rectangle at the ground level drawn in a two-point perspective. In the sequence of Figures 5.1.38 through 5.1.40 and the associated captions we describe a method for subdividing a line segment (a side of the rectangle) into three line segments of equal length.
   a. Use this method to subdivide the other two sides of the rectangle in Figure 5.1.37 into three equal parts.
   b. Use this method to construct a perspective image of a five-by-five chessboard over the given rectangle.

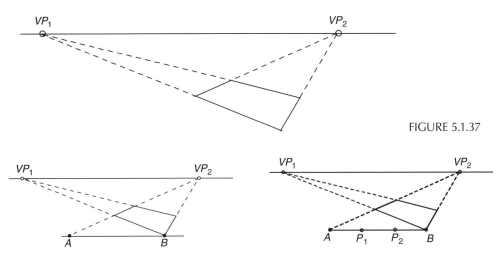

FIGURE 5.1.37

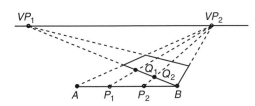

FIGURE 5.1.38 We draw a line parallel to the horizon and identify the line segment AB as shown.

FIGURE 5.1.39 Use Construction 7 in Section 1.2 to subdivide AB into three equal parts (the subdividing points P_1 and P_2 are indicated).

FIGURE 5.1.40 Join the division points with vanishing points as shown. The intersection of these lines with a pair of parallel sides of the rectangle yields the points of subdivision Q_1 and Q_2.

9*. The picture in Figure 5.1.41 shows a perspective scheme of a (horizontal) railroad track with the two parallel rails depicted by the two intersecting lines, and two consecutive

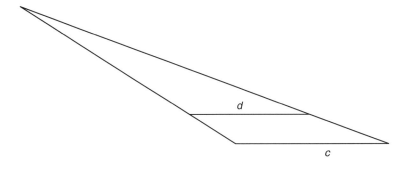

FIGURE 5.1.41

parallel wooden railroad ties represented by the parallel line segments denoted by c and d. Assume that any two consecutive railroad ties are placed at the same fixed distance. Precisely, draw a few more railroad ties. [*Hint*: It might be a good idea to first draw a nonperspective scheme of the railroad, with the railroad tracks depicted as parallel and the railroad ties also depicted as parallel. Then observe that some diagonals of any two rectangles made of the ties and tracks are also parallel.]

5.2* MATHEMATICS OF PERSPECTIVE DRAWING: A BRIEF OVERVIEW (OPTIONAL)

Introduction

From the mathematical point of view (disregarding colors, intentional violations of the rules, and other elements of artistic choice), a perspective drawing is completely determined by the scene on one side of the drawing plane, and the position of the point of view (*the **eye position***) on the other side of the drawing plane. The *scene* we draw is then on the other side of the drawing plane. The perspective drawing associated with the given choice of the drawing plane and the eye position consists of determining the intersections between the drawing plane and the line segments starting at the eye and ending at the points in the scene (see Figure 5.2.1).

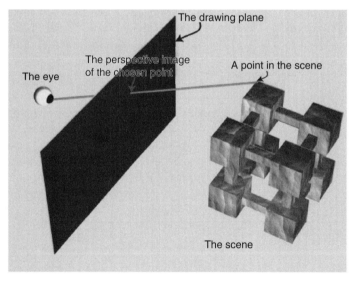

FIGURE 5.2.1 □

A Formula for Perspective Images of Points

First we recall the three-dimensional rectangular coordinate system. It consists of three intersecting, oriented, graded, and mutually perpendicular lines: the x-axis, the y-axis, and the z-axis. The first two form the two-dimensional coordinate system for the xy-plane; we have encountered them in Chapter 2. The points in three dimensions now correspond to triples of numbers. The additional z-axis gives rise to the third coordinate that measures the vertical distances to the *xy*-plane containing the *x* and *y* axes: it is positive if the point is above the *xy*-plane, and negative if it is below this plane. In Figure 5.2.2, we show a point *A* and we see that its coordinates *a*, *b*, and *c* correspond to certain distances as indicated. Each point in the three-dimensional space with a coordinate system is uniquely determined by its three coordinates.

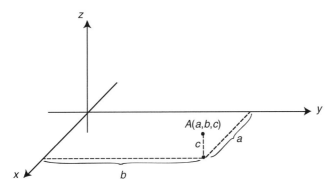

FIGURE 5.2.2

We will now show how to find the coordinates of the perspective image in the drawing plane of every point in the scene if we are given the coordinates of the eye and the position of the drawing plane. In order to simplify the argument, we will assume that the eye position is at $(0, 0, e)$, where *e* is any positive number larger than 1 (so that the eye position is above the drawing plane; see Figure 5.2.3). Further, we assume that the drawing plane is parallel to the *xy*-plane, and 1 unit above it, so that its equation is $z = 1$ (indicating that the third coordinates of all the points in the drawing plane are equal to 1). These assumptions are not restrictive since we can always position the coordinate system in such a way that the eye position and the position of the drawing plane are as indicated above. Our choice

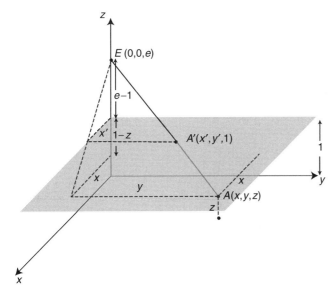

FIGURE 5.2.3

of a horizontal drawing plane is a compromise between the standard mathematical view of the coordinate axis (with the z-axis vertical) and the setting of the coordinate axes that computer graphic artists use (where the z-axis is horizontal while the y-axis is vertical). We keep the standard mathematical view on the axes, but use the drawing plane that would have been vertical in the computer graphics setting. The substance, of course, remains the same in both settings.

Our goal is to find the coordinates of the perspective image A' of the point A. The third coordinate is equal to 1 simply because A' lies on the drawing plane.

Now we focus on x'. Consider the two right-angled triangles shown in the xz-plane: one has sides x', $e - 1$ and top vertex E, while the larger triangle has the sides x, $(e - 1) + (1 - z) = e - z$ and the same top vertex E (see Figure 5.2.3). It follows from the similarity of these two triangles that $\frac{x'}{x} = \frac{e-1}{e-z}$, and so $x' = \frac{x(e-1)}{e-z}$.

A completely symmetric argument yields a similar formula for y': $y' = \frac{y(e-1)}{e-z}$.

So, we can summarize our argument in the following formula for perspective images of points.

> If the drawing plane is horizontal and 1 unit above the xy-plane and if the eye position is at $(0, 0, e)$, then the perspective image of a point A with coordinates (x, y, z) is a point A' with coordinates $\left(\frac{x(e-1)}{e-z}, \frac{y(e-1)}{e-z}, 1 \right)$. □

Basic Example 1

Find the coordinates of the perspective image A' of the point $A(1, 2, -1)$ if the eye position is at $E(0, 0, 4)$ and the drawing plane is as above.

Solution. We simply substitute $e = 4$, $x = 1$, $y = 2$, and $z = -1$ in the above formula to get $\left(\frac{x(e-1)}{e-z}, \frac{y(e-1)}{e-z}, 1 \right) = \left(\frac{3}{5}, \frac{6}{5}, 1 \right)$ for the coordinates of A'. □

Basic Example 2

Find the equation of the perspective image of the x-axis if the eye position is at $E(0, 0, e)$, $e > 1$, and the drawing plane is $z = 1$ (as above).

Solution. First note that, under our setting, the x-axis is completely within the scene (i.e., it is in the half plane on the side of the drawing plane not containing the eye position). Points on the x-axis have the second and the third coordinates equal to 0, while the first coordinates vary through the set of all (real) numbers. So, points on this line have coordinates of type $(x, 0, 0)$. Consequently, and according to our formula, the perspective images of a point of this type will have coordinates $\left(\frac{x(e-1)}{e-0}, \frac{0(e-1)}{e-0}, 1 \right) = \left(x\left(1 - \frac{1}{e}\right), 0, 1 \right)$, where e is a fixed number larger than 1 and x ranges through all numbers (see Figure 5.2.4).

Since we have found the set of perspective images of the points on the x-axis, we have accomplished our goal and we may, if we wish, stop at this point. However, let us go one

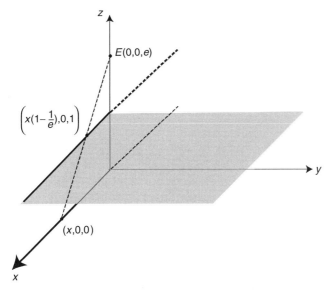

FIGURE 5.2.4

step further and analyze this set. As x ranges through all numbers, so does any nonzero multiple of x, and so $x\left(1 - \frac{1}{e}\right)$ also ranges through the set of all numbers. Consequently, the image points $\left(x\left(1 - \frac{1}{e}\right), 0, 1\right)$ under the perspective mapping have first coordinates that assume all values, whereas the second must be 0 and the third 1. It follows that the set of all points $\left(x\left(1 - \frac{1}{e}\right), 0, 1\right)$ is the line that is parallel to the x-axis and 1 unit above it (see Figure 5.2.4). $\qquad\square$

As we saw in the last exercise, the perspective image of a line (the x-axis) is again a line. The x-axis is a special line: it is parallel to the drawing plane, and it is completely within the scene. The lines that are not parallel to the drawing plane will only partially be within the scene. The perspective images of semilines within the scene are semilines in the drawing plane. An equivalent statement is given in the following principle.

A Perspective Principle

Perspective Principle: If three points in the scene lie on a straight line then their images also lie on a single line in the drawing plane.

Proof. We provide a geometric (visual) justification.

If the three points in the scene are on a line that passes through the eye position, then they will all have a single point as their image in the drawing plane (which, of course, will be on some line).

Now consider the generic case when the line containing the three points does not pass through the eye position. Then, this line and the point at the eye position determine a single plane Σ that contains them (see Figure 5.2.5). Since the line segment connecting the point at the eye position with any of these three points is in Σ, it follows that the perspective projection of this point will also be in Σ. So, the perspective images of all these three points are in Σ. Since the perspective images of the points are in the drawing plane, it follows that

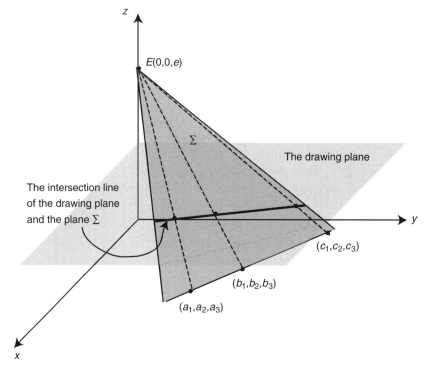

FIGURE 5.2.5

the perspective images of these three points are both in the drawing plane and in Σ. The intersection of these two planes is a line, and so it follows that the perspective images of the three points are on a single line.

It follows from the perspective principle that perspective images of lines are lines. ☐

Homogeneous Coordinates and The Matrix Form of the Formula for Perspective Images of Points

We will now introduce matrices in relation to perspective drawing. The main advantage in doing so is in the elegance of expressing what we already know, in the potential of using this elegance to continue developing the theory, and in providing a route of encoding perspective drawing in, say, computer graphics. We only present the main idea of the theory.

Suppose A is a point with coordinates (u, v, w) in the three-dimensional coordinate system. ***Homogeneous coordinates*** of the point A are the coordinates of any quadruple (a, b, c, d) (where $d \neq 0$) such that $\frac{a}{d} = u$, $\frac{b}{d} = v$ and $\frac{c}{d} = w$ [so that $\left(\frac{a}{d}, \frac{b}{d}, \frac{c}{d}\right)$ are the coordinates of the point A]. Note again that the point A defined by the quadruple (a, b, c, d) lies in three dimensions—it is not a point in four dimensions as the number of coordinates may suggest. Given a quadruple (a, b, c, d) of homogeneous coordinates we denote the point

$\left(\dfrac{a}{d}, \dfrac{b}{d}, \dfrac{c}{d}\right)$ defined by the quadruple $h(a, b, c, d)$. For example, $(3, 6, 9, 3)$ and $(1, 2, 3, 1)$ are both homogeneous coordinates of the point with coordinates $(1, 2, 3)$. Alternatively, using the notation we have just introduced, we have $h(3, 6, 9, 3) = h(1, 2, 3, 1) = (1, 2, 3)$. As we see, the coordinates in each quadruple (a, b, c, d), $d \neq 0$, are homogeneous coordinates of a unique point A in the space, but a point in the space has infinitely many quadruples of homogeneous coordinates: each quadruple (ta, tb, tc, td), $t \neq 0$, defines the homogeneous coordinates of the same point A.

We will now introduce a matrix corresponding to the horizontal drawing plane being used here, and to the eye position at $(0, 0, e)$. The reason we choose this particular matrix will be made clear in a moment.

$$P = \begin{bmatrix} e-1 & 0 & 0 & 0 \\ 0 & e-1 & 0 & 0 \\ 0 & 0 & -1 & e \\ 0 & 0 & -1 & e \end{bmatrix}$$

Call this matrix P the **matrix of our perspective.**

Recall the product formula $\begin{bmatrix} a_1 & a_2 \\ b_1 & b_2 \end{bmatrix} \begin{bmatrix} x_1 \\ x_2 \end{bmatrix} = \begin{bmatrix} a_1 x_1 + a_2 x_2 \\ b_1 x_1 + b_2 x_2 \end{bmatrix}$ defined in Section 2.2. The

product formula for 3×3 matrices is similar: $\begin{bmatrix} a_1 & a_2 & a_3 \\ b_1 & b_2 & b_3 \\ c_1 & c_2 & c_3 \end{bmatrix} \begin{bmatrix} x_1 \\ x_2 \\ x_3 \end{bmatrix} = \begin{bmatrix} a_1 x_1 + a_2 x_2 + a_3 x_3 \\ b_1 x_1 + b_2 x_2 + b_3 x_3 \\ c_1 x_1 + c_2 x_2 + c_3 x_3 \end{bmatrix}$, as is

the formula for 4×4 matrices: $\begin{bmatrix} a_1 & a_2 & a_3 & a_4 \\ b_1 & b_2 & b_3 & b_4 \\ c_1 & c_2 & c_3 & c_4 \\ d_1 & d_2 & d_3 & d_4 \end{bmatrix} \begin{bmatrix} x_1 \\ x_2 \\ x_3 \\ x_4 \end{bmatrix} = \begin{bmatrix} a_1 x_1 + a_2 x_2 + a_3 x_3 + a_4 x_4 \\ b_1 x_1 + b_2 x_2 + b_3 x_3 + b_4 x_4 \\ c_1 x_1 + c_2 x_2 + c_3 x_3 + c_4 x_4 \\ d_1 x_1 + d_2 x_2 + d_3 x_3 + d_4 x_4 \end{bmatrix}.$

Now, take any point (x, y, z) and the quadruple $(x, y, z, 1)$ giving homogeneous coordinates of this point, and compute the product:

$$\begin{bmatrix} e-1 & 0 & 0 & 0 \\ 0 & e-1 & 0 & 0 \\ 0 & 0 & -1 & e \\ 0 & 0 & -1 & e \end{bmatrix} \cdot \begin{bmatrix} x \\ y \\ z \\ 1 \end{bmatrix} = \begin{bmatrix} x(e-1) \\ y(e-1) \\ e-z \\ e-z \end{bmatrix}$$

The quadruple that is the result of the product can again be considered as giving homogeneous coordinates of a point in the space, in which case $h(x(e-1), y(e-1), e-z, e-z) = \left(\dfrac{x(e-1)}{e-z}, \dfrac{y(e-1)}{e-z}, \dfrac{e-z}{e-z}\right) = \left(\dfrac{x(e-1)}{e-z}, \dfrac{y(e-1)}{e-z}, 1\right)$, which is exactly the perspective projection of the point (x, y, z). We note that this formula makes sense only if $e \neq z$, that is, only if the point (x, y, z) is not in the horizontal plane containing the eye position.

We have merely made it possible to restate the formula for perspective images of points using the language of matrices:

If the drawing plane is horizontal and 1 unit above the xy-plane, and if the eye position is at $(0, 0, e)$, with $e > 1$, then the perspective image of a point A with coordinates (x, y, z) is a point A' with coordinates $h\begin{pmatrix} \begin{bmatrix} e-1 & 0 & 0 & 0 \\ 0 & e-1 & 0 & 0 \\ 0 & 0 & -1 & e \\ 0 & 0 & -1 & e \end{bmatrix} \cdot \begin{bmatrix} x \\ y \\ z \\ 1 \end{bmatrix} \end{pmatrix}$.

So, the matrix we have introduced contains all the data needed to establish projective images of points under our setting. We point out again that the main advantage of using the theory of matrices is that its simplicity and elegance make it easier to deal with more general or more complicated cases (say, when the drawing plane is not horizontal). □

Example

Consider the cube with vertices $(1, 1, 0)$, $(2, 1, 0)$, $(2, 2, 0)$, $(1, 2, 0)$ (top face), and $(1, 1, -1)$, $(2, 1, -1)$, $(2, 2, -1)$, $(1, 2, -1)$ (bottom face). Use $(0, 0, 3)$ for the eye position and the horizontal drawing plane $z = 1$ we employed above, to find the perspective images of the edges of this cube and plot them in the drawing plane.

Solution. We compute the perspective images of the eight vertices:

For the point $(1, 1, 0)$: $h\begin{pmatrix} \begin{bmatrix} 3-1 & 0 & 0 & 0 \\ 0 & 3-1 & 0 & 0 \\ 0 & 0 & -1 & 3 \\ 0 & 0 & -1 & 3 \end{bmatrix} \cdot \begin{bmatrix} 1 \\ 1 \\ 0 \\ 1 \end{bmatrix} \end{pmatrix} = h\begin{bmatrix} 2 \\ 2 \\ 3 \\ 3 \end{bmatrix} = \left(\frac{2}{3}, \frac{2}{3}, 1\right)$

For the point $(2, 1, 0)$: $h\begin{pmatrix} \begin{bmatrix} 2 & 0 & 0 & 0 \\ 0 & 2 & 0 & 0 \\ 0 & 0 & -1 & 3 \\ 0 & 0 & -1 & 3 \end{bmatrix} \cdot \begin{bmatrix} 2 \\ 1 \\ 0 \\ 1 \end{bmatrix} \end{pmatrix} = h\begin{bmatrix} 4 \\ 2 \\ 3 \\ 3 \end{bmatrix} = \left(\frac{4}{3}, \frac{2}{3}, 1\right)$

For the point $(2, 2, 0)$: $h\begin{pmatrix} \begin{bmatrix} 2 & 0 & 0 & 0 \\ 0 & 2 & 0 & 0 \\ 0 & 0 & -1 & 3 \\ 0 & 0 & -1 & 3 \end{bmatrix} \cdot \begin{bmatrix} 2 \\ 2 \\ 0 \\ 1 \end{bmatrix} \end{pmatrix} = h\begin{bmatrix} 4 \\ 4 \\ 3 \\ 3 \end{bmatrix} = \left(\frac{4}{3}, \frac{4}{3}, 1\right)$

For the point $(1, 2, 0)$: $h\begin{pmatrix} \begin{bmatrix} 2 & 0 & 0 & 0 \\ 0 & 2 & 0 & 0 \\ 0 & 0 & -1 & 3 \\ 0 & 0 & -1 & 3 \end{bmatrix} \cdot \begin{bmatrix} 1 \\ 2 \\ 0 \\ 1 \end{bmatrix} \end{pmatrix} = h\begin{bmatrix} 2 \\ 4 \\ 3 \\ 3 \end{bmatrix} = \left(\frac{2}{3}, \frac{4}{3}, 1\right)$

For the point $(1, 1, -1)$: $h\begin{pmatrix} \begin{bmatrix} 3-1 & 0 & 0 & 0 \\ 0 & 3-1 & 0 & 0 \\ 0 & 0 & -1 & 3 \\ 0 & 0 & -1 & 3 \end{bmatrix} \cdot \begin{bmatrix} 1 \\ 1 \\ -1 \\ 1 \end{bmatrix} \end{pmatrix} = h\begin{bmatrix} 2 \\ 2 \\ 4 \\ 4 \end{bmatrix} = \left(\frac{2}{4}, \frac{2}{4}, 1\right) = \left(\frac{1}{2}, \frac{1}{2}, 1\right)$

For the point $(2, 1, -1)$: $h\left(\begin{bmatrix} 2 & 0 & 0 & 0 \\ 0 & 2 & 0 & 0 \\ 0 & 0 & -1 & 3 \\ 0 & 0 & -1 & 3 \end{bmatrix} \cdot \begin{bmatrix} 2 \\ 1 \\ -1 \\ 1 \end{bmatrix}\right) = h\left(\begin{bmatrix} 4 \\ 2 \\ 4 \\ 4 \end{bmatrix}\right) = \left(\frac{4}{4}, \frac{2}{4}, 1\right) = \left(1, \frac{1}{2}, 1\right)$

For the point $(2, 2, -1)$: $h\left(\begin{bmatrix} 2 & 0 & 0 & 0 \\ 0 & 2 & 0 & 0 \\ 0 & 0 & -1 & 3 \\ 0 & 0 & -1 & 3 \end{bmatrix} \cdot \begin{bmatrix} 2 \\ 2 \\ -1 \\ 1 \end{bmatrix}\right) = h\left(\begin{bmatrix} 4 \\ 4 \\ 4 \\ 4 \end{bmatrix}\right) = (1, 1, 1)$

For the point $(1, 2, -1)$: $h\left(\begin{bmatrix} 2 & 0 & 0 & 0 \\ 0 & 2 & 0 & 0 \\ 0 & 0 & -1 & 3 \\ 0 & 0 & -1 & 3 \end{bmatrix} \cdot \begin{bmatrix} 1 \\ 2 \\ -1 \\ 1 \end{bmatrix}\right) = h\left(\begin{bmatrix} 2 \\ 4 \\ 4 \\ 4 \end{bmatrix}\right) = \left(\frac{2}{4}, \frac{4}{4}, 1\right) = \left(\frac{1}{2}, 1, 1\right)$

We provide in Figure 5.2.6, a computer-generated plot of these eight points in the drawing plane together with the joining edges. We have used the Perspective Principle (stated earlier in this section) to justify the implicit claim that the edges of the cube are projected to line segments.

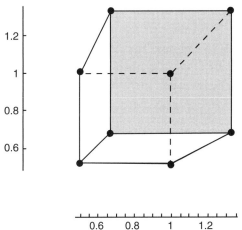

FIGURE 5.2.6 □

Note the relative simplicity of the above construction of the perspective image of a perfect cube, and compare it with the more involved construction of a perspective image of a square given in Section 5.1. It is clear that the small theory we are developing here has some advantages with respect to the traditional approach.

Rule 2. A Mathematical Approach

Recall Rule 2 from Section 5.1.

Lines parallel to the drawing plane are depicted as parallel lines in perspective drawing.

We will use our theory to give an algebraic proof of that claim.

Proof. Keep the setting of the eye position and the drawing plane as above. Since we are assuming that pairs of parallel lines are parallel to the drawing plane, it follows that these two lines are also horizontal, and so we may assume (without significant loss of generality) that they are in the xy-plane. We will also assume that these two lines are not parallel to the y-axis, so that their slopes within the xy-plane are finite. Since they are parallel they have the same slope, and so, their equations are of the form $y = mx + b_1$ (for the first line) and $y = mx + b_2$ (for the second line); the third coordinate is $z = 0$. So, the points on the first line have coordinates $(x, y, z) = (x, mx + b_1, 0)$ and the points on the second line have coordinates $(x, y, z) = (x, mx + b_2, 0)$. We now find the perspective images of the points of the first line using homogeneous coordinates and the matrix of our perspective:

$$\begin{bmatrix} e-1 & 0 & 0 & 0 \\ 0 & e-1 & 0 & 0 \\ 0 & 0 & -1 & e \\ 0 & 0 & -1 & e \end{bmatrix} \cdot \begin{bmatrix} x \\ mx+b_1 \\ 0 \\ 1 \end{bmatrix} = \begin{bmatrix} x(e-1) \\ (mx+b_1)(e-1) \\ e \\ e \end{bmatrix}$$

So, the coordinates of the perspective images of the points of the first line are $\left(\dfrac{x(e-1)}{e}, \dfrac{(mx+b_1)(e-1)}{e}, 1 \right)$. Focusing on the second coordinate, we see that $\dfrac{(mx+b_1)(e-1)}{e} = m\dfrac{x(e-1)}{e} + \dfrac{b_1(e-1)}{e}$. Denoting $x' = \dfrac{x(e-1)}{e}$, $y' = \dfrac{(mx+b_1)(e-1)}{e}$, this gives that $y' = mx' + \dfrac{b_1(e-1)}{e}$. The equation $y' = mx' + \dfrac{b_1(e-1)}{e}$ determines a line with slope m. So we have proved more than we claimed: the perspective image of a line with a slope m in the xy-plane is always a line of the same slope in the drawing plane. Consequently, since the slopes of the pair of parallel lines do not change (only their horizontal plane changes), perspective images of parallel lines are indeed parallel. □

The other rules of perspective drawing that we have stated in Section 5.1 can now be proven precisely within the setting of the above theory. Moreover, our theory makes it possible to expand on these rules and prove other properties of perspective drawing. For example, it is possible to show within our theory that perspective images of circles are ovals (more accurately, they are ellipses: ellipses will be discussed in Section 5.4).

Summarizing, perspective images of lines are lines, parallel lines may or may not be mapped to parallel lines, circles in general get distorted into ovals, and, of course, objects far away from the drawing plane are depicted smaller than objects of the same size that are closer to the drawing plane. We illustrate all these rules with Figures 5.2.7 and 5.2.8.

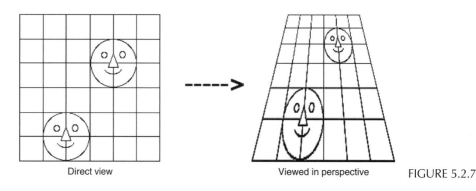

Direct view   Viewed in perspective   FIGURE 5.2.7

The two pictures in Figure 5.2.7 are set in three dimensions in Figure 5.2.8.

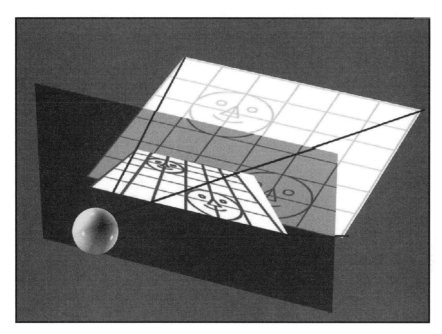

FIGURE 5.2.8

The fact that perspective images of circles are ellipses is used cleverly in the piece of art depicting simple rows and columns of circles and ellipses (Figure 5.2.9). Our perception of a three-dimensional wavy pattern is a consequence of our acquired inclination to extract or decode three-dimensional scenes from two-dimensional pictures.

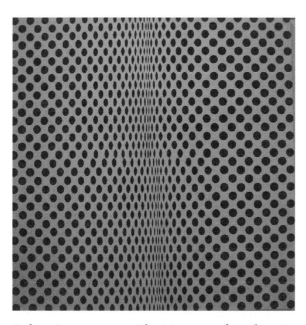

FIGURE 5.2.9 Bridget Riley, *Current*, 1964, The Museum of Modern Art, New York.

A Formula for the Vanishing Points

We will now deduce a formula for the coordinates of the vanishing point corresponding to a class of lines parallel to a given line. We keep the same setting as above: a horizontal drawing plane at $z = 1$ and the eye position at $(0, 0, e)$ above the drawing plane (so, $e > 1$). In order to avoid lengthy arguments we will not be very formal; nevertheless, the formula we will derive will be precise and useful. A warning to the reader: at one point in our argument, we will use basic linear algebra to get the parametric equations of a line passing through a given point and parallel to a given vector (recall that vectors are oriented line segments).

We start with a line l intersecting the drawing plane at the point C. We draw the line through the eye position and parallel to l, and we denote by V the intersection of this line and the drawing plane (see Figure 5.2.10). We saw above (see Figure 5.2.5) that the perspective image of the line l is in the intersection of the drawing plane and the plane through E and l. Since the line EV is parallel to l, it follows from the definition of parallel lines in three dimensions that this line is also in the plane through E and l. So, the point V, being both in the drawing plane and in the plane through E and l, must be a point in the intersection of these two planes, and so it is on the projection of the line l in the drawing plane.

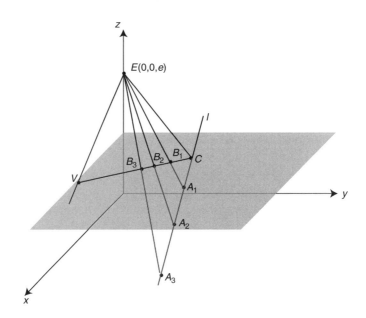

FIGURE 5.2.10 We show three points, A_i, $i = 1, 2, 3$ and their perspective images B_i, $i = 1, 2, 3$ on the line CV.

Notice again that V is the intersection point of the drawing plane and the line through E and parallel to l. Hence, if we take any other line m parallel to l, then the line through E and parallel to m is the same as the line through E and parallel to l, and so the point V is in the perspective image of m as well (see Figure 5.2.11). Consequently, the perspective images of every line parallel to the line l must pass through the point V, and thereby V is the vanishing point corresponding to the class of lines that are parallel to l.

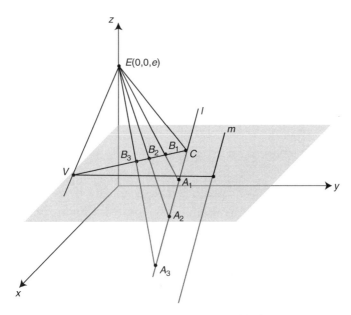

FIGURE 5.2.11 Another line m that is parallel to l will yield the same point V.

In order to derive the formula for the coordinates of V, we need to borrow the following equations from basic linear algebra. If a line passes through a point with coordinates (a, b, c) and is parallel to a vector starting at the origin and ending at a point with coordinates (v_1, v_2, v_3), then the coordinates (x, y, z) of the points on the line satisfy the following **parametric equations** of the line: $x = v_1 t + a$, $y = v_2 t + b$, $z = v_3 t + c$, where the *parameter t* varies through all real numbers.

Now, suppose that the line l is parallel to a vector (v_1, v_2, v_3), where a vector denoted by (v_1, v_2, v_3) is the one starting at the origin and ending at the point (v_1, v_2, v_3). Then the ray from E to V, being parallel to l, is also parallel to the vector (v_1, v_2, v_3). Since the coordinates of E are $(0, 0, e)$, the parametric equations of the line through E and V are $x = v_1 t + 0$, $y = v_2 t + 0$, $z = v_3 t + e$. Simplifying a bit, we get $x = v_1 t$, $y = v_2 t$, $z = v_3 t + e$. The line l was chosen so that it is not parallel to or on the drawing plane: so, it must be that $v_3 \neq 0$, or else the vector (v_1, v_2, v_3) would be in the xy-plane and so parallel to the drawing plane. We conclude that $x = v_1 t$, $y = v_2 t$, $z = v_3 t + e$, $v_3 \neq 0$, are parametric equations for the line through E and V.

The point V is in the intersection of the drawing plane and the ray passing through E and V, so, to complete our search, we need to find the coordinates of the intersection of this ray (defined by $x = v_1 t, y = v_2 t, z = v_3 t + e$, where t changes through all numbers) and the drawing plane (which, as we have seen, is defined by $z = 1$). This amounts to solving all the equations $x = v_1 t, y = v_2 t, z = v_3 t + e$, and $z = 1$ simultaneously. It follows from $z = v_3 t + e$, and $z = 1$ that $v_3 t + e = 1$, from where we find that $t = \dfrac{1-e}{v_3}$ (note that it is precisely here that we need $v_3 \neq 0$). Substituting this into $x = v_1 t, y = v_2 t$, we get $x = v_1 \dfrac{1-e}{v_3}$ and $y = v_2 \dfrac{1-e}{v_3}$.

That completes our search; the point V has coordinates $\left(v_1 \dfrac{1-e}{v_3}, v_2 \dfrac{1-e}{v_3}, 1\right)$.

We summarize: The vanishing point corresponding to all lines parallel to the vector (v_1, v_2, v_3), where $v_3 \neq 0$, has coordinates $\left(v_1 \dfrac{1-e}{v_3}, v_2 \dfrac{1-e}{v_3}, 1\right)$. □

In Figure 5.2.12, we show one piece of an artwork: it is a depiction of the Pantheon on a spherical canvas. The unusual shape of the canvas was used to obtain a perspective image with six vanishing points. More details are given in the caption of Figure 5.2.12.

FIGURE 5.2.12 **(See color insert following page 144.)** Dick Termes, *Pantheon*, 13″ sphere, 1998. Here is how Dick Termes describes the six-point perspective: "Six point perspective shows up when you study what happens to the lines that create a cubical room. When you stand inside that room you will notice there are three sets of parallel lines that create that room. If you look closely at those lines one set of those lines vanishes to the north and if you turn you will see that same set vanishes to the south. The next set vanishes off to the east and the same set vanishes off to the west. The last set of lines of that room project up directly above your head and the same lines project down below your feet. That is six point perspective. When you draw this on the sphere it is much more obvious."

Exercises: (In all the problems below, the drawing plane is as above, at $z = 1$.)
  1. a. The quadruples (1, 4, 2, 4) and (−2, −6, −4, 2) are homogeneous coordinates of two points in a three-dimensional coordinate system. Identify the coordinates of these points and plot the points in the three-dimensional coordinate system.
     b. Find the perspective images of the two points in part (a) if the eye position is at (0, 0, 2). Plot the perspective images of these two points in a three-dimensional coordinate system.
  2. Find the perspective image of a square with vertices at (0, 0, 0), (1, 0, 0), (1, 1, 0), and (0, 1, 0) if the eye position is at (0, 0, 4). Make a two-dimensional perspective sketch of the square in the drawing plane.
  3. Find the perspective image of a cube with vertices at (0, 0, 0), (2, 0, 0), (2, 2, 0), (0, 2, 0) (top), and (0, 0, −2), (2, 0, −2), (2, 2, −2), (0, 2, −2) (bottom) if the eye position is at (0, 0, 2). Make a two-dimensional sketch in the drawing plane (as in Figure 5.2.6).
  4. Find the perspective image of a square pyramid with base vertices at (2, 2, 0), (0, 2, 0), (2, 2, −2), (0, 2, −2) and the top apex at (1, 4, −1) if the eye position is at (0, 0, 2).
  5. Find the coordinates of a vanishing point corresponding to the class of lines parallel to the vector (1, 2, 3) if the eye position is at (0, 0, 3).

6. Find the coordinates of the vanishing point corresponding to the class of lines parallel to the line passing through the origin and the point (1, 1, 1) if the eye position is at (0, 0, 2).

7. Find the coordinates of the vanishing point corresponding to the class of lines parallel to the line passing through the points (−2, 1, −3) and (1, 3, −4) if the eye position is at (0, 0, 2). [*Hint*: you first need to find the coordinates of a point *A* such that the vector from the origin to *A* is parallel to the vector from (−2, 1, −3) to (1, 3, −4).]

5.3 REGULAR AND OTHER POLYHEDRA

In two dimensions, there are infinitely many types of regular polygons: equilateral triangle, square, regular pentagon, regular hexagon, etc.; we can have a regular polygon of as many sides as we wish. As we will see in this section, the situation in three dimensions is different. Before we see why this is so, we need to define the three-dimensional analogs of regular polygons.

Regular Polyhedra (Platonic Solids)

We have mentioned convex objects in Chapter 2. Recall that a geometric object is **convex** if for any two points in that object the line segment between the two points is entirely within the object. For example, a solid cube is convex, but an empty box is not convex. Unless otherwise stated, all of the geometric objects that we consider in this section are convex.

A **polyhedron** is a bounded three-dimensional object whose boundary is made of (filled) polygons (called the **faces** of the polyhedron). We assume here and throughout that any two bounding polygons have either one common edge (with the two end vertices), or one common vertex (corner), or nothing common at all. For example, two adjacent polygons could not share only one half of an edge. If a polyhedron is convex, then it must be *filled*; that is, every point in the interior of a polyhedron is considered to be a part of that polyhedron. (Filled) cubes and any (filled) boxes are filled polyhedra. Filled polyhedra are sometimes called *solids*. From now on, we assume that every polyhedron is filled.

We have to be careful in our definition of the three-dimensional equivalent of a regular polygon: we would like this definition to yield a *completely symmetric* object. For example, it would *not* be sufficient to stipulate that all of the bounding polygons are congruent (equal in size and shape) regular polygons. A counterexample is shown in Figure 5.3.1.

FIGURE 5.3.1 This is a (so-called) triangular dipyramid (we will see much longer names of polyhedra further in this section): all of the sides are congruent equilateral triangles. However, the vertices of this polyhedron are not *symmetric*: notice that both the top and the bottom vertices are adjacent to three triangles, whereas each of the middle vertices is adjacent to four triangles.

As we see, we need an extra condition to guarantee that the neighborhoods of the vertices of the bounding polygons are of the same kind.

A convex polyhedron is *regular* if all the bounding polygons are congruent regular polygons and if each vertex of a bounding polygon is adjacent to the same number of bounding polygons. Regular polyhedra are also called *Platonic solids*. The Greek philosopher Plato analyzed them and used them prominently in his philosophical systems of the universe.

How many regular polyhedra are there? Very few! As known from antiquity, there are only five such objects (see Figures 5.3.2 through 5.3.6 and remember that what you see are perspective images of three-dimensional objects). The longish names are derived from Greek words, and the first part of these words is simply the number of faces of the solid; for example, *dodeca* (see dodecahedron in Figure 5.3.5) is 12 in Greek.

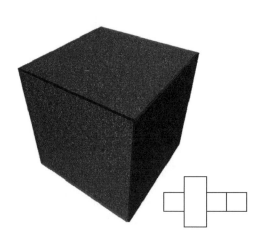

FIGURE 5.3.2 Cube: six square sides.

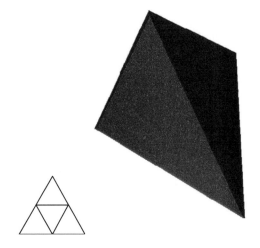

FIGURE 5.3.3 Tetrahedron: the simplest of Platonic solids. Its four sides are all equilateral triangles.

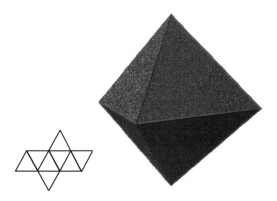

FIGURE 5.3.4 Octahedron: eight equilateral triangles.

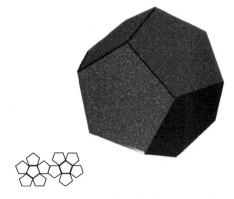

FIGURE 5.3.5 Dodecahedron: twelve regular pentagons.

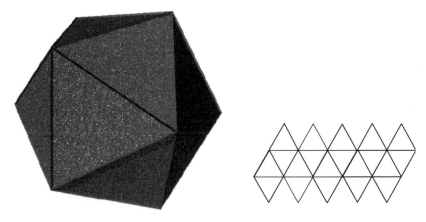

FIGURE 5.3.6 Icosahedron: twenty equilateral triangles.

A Bit of Math. There are Only Five Regular Polyhedra
We will now show that there are exactly five regular polyhedra. The earliest preserved written proof of this fact was given by Euclid in the last proposition of his *Elements*.

Every vertex V of a regular polyhedron is surrounded by a number of regular polygons of the same type and size. So, the interior angles of the polygons around V are all equal—denote them by α. The sum of these angles must be less than 360°. In order to see this just spread all the polygons around V over one plane (without detaching them from V and without altering their shape). If the number of polygons around each vertex is n, then the preceding claim can be expressed by the inequality $n\alpha < 360°$, or equivalently, $\alpha < \frac{360°}{n}$. Since there must be at least three polygons around each vertex in a polyhedron, we have that $n \geq 3$. So, $\frac{360°}{n} \leq \frac{360°}{3}$. The inequalities $\alpha < \frac{360°}{n}$ and $\frac{360°}{n} \leq \frac{360°}{3}$ imply that $\alpha < \frac{360°}{3}$, or $\alpha < 120°$. So, whatever regular polygon is used, its interior angle must be less than 120°. This excludes hexagons (with interior angles equal to 120°) and regular polygons with the number of sides larger than six (since they have interior angles larger than 120°). We are left with equilateral triangles, squares, and regular pentagons. If a vertex in a regular polyhedron has four or more adjacent regular pentagons, then $n\alpha \geq 4\alpha$, where α is the interior angle of a regular pentagon. But for a regular pentagon we have that $\alpha = 108°$, and so $n\alpha \geq (4)(108°) = 432°$. As we have already noticed above, $n\alpha < 360°$, and so a vertex cannot have four or more adjacent regular pentagons. That leaves having three pentagons around each vertex as the only possibility of a regular polyhedron with faces being made of regular pentagons (the dodecahedron shown above). Similar analysis shows that we have only one regular polyhedron with square faces (the cube), and three regular polyhedra with triangular faces (tetrahedron, octahedron, and icosahedron). □

As we have pointed out above, Plato produced one of the first written records on regular polyhedra which survived until the present time. He was so impressed by their elegance that he used them as symbols representing various attributes of the universe. He writes in his dialogue *Timaeus* (fourth century BC): "Let us assign the cube to earth, for it is the most immobile of the four bodies and most retentive of shape." Thus, the earth, being the least mobile, was represented in Plato's scheme by a cube, and then, in the order given by their mobility, water was represented by an icosahedron, air by an octahedron, and, fire, the most mobile of them, by a tetrahedron. According to Plato, God used the fifth regular polyhedron (the dodecahedron) "for embroidering the constellations of the whole heaven."

FIGURE 5.3.7 *Neolitic stone models of regular polyhedra*, Ashmolean Museum of Art and Archeology, second millennium BC or earlier. (Some archeologists have dated these stone polyhedra to as early as 20000 BC.)

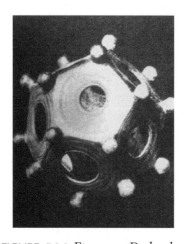

FIGURE 5.3.8 Etruscan *Dodecahedron*, second century BC or earlier.

Plato was not the discoverer of the regular polyhedra. Pythagoreans before him considered these objects in their work on geometry. Moreover, various archeological findings imply that other ancient civilizations were also aware of the regular polyhedra . For example (see Emmer, p. 215), archeologists found an Etruscan dodecahedron from the fifth century BC or earlier near Padua in Italy. Further, the Neolithic people of Scotland of the second millennium BC apparently knew about the regular solids. Figure 5.3.7 shows the Scottish stone models of various polyhedra. Figure 5.3.8 shows a second century BC dodecahedron excavated in Germany.

Semiregular Polyhedra (Archimedean Solids)

We now relax one of the conditions in the definition of regular polyhedra. Suppose we do not require that the faces are congruent copies of a single type of a regular polygon. Polyhedra satisfying all the conditions of regular polyhedra except the one just mentioned are called semiregular polyhedra. Explicitly, a convex, bounded polyhedron is **semiregular** if all the bounding polygons are regular polygons (possibly more than one type), all of them with edges of equal length, if each vertex of a bounding polygon is adjacent to the same number of polygons, and finally, if there exists a fixed cyclic order of the types of polygons around all the vertices.

The number of types of semiregular polyhedra is much larger than the number of regular polyhedra—in fact, there are infinitely many different (in shape) semiregular polyhedra. Can you see why?

The main reason for the abundance of semiregular polyhedra is that they include prisms (with a regular polygon with as many edges as we require as their bottom and top faces). Figure 5.3.9 shows a hexagonal prism: it has hexagons at the top and the bottom, while the side faces are all squares.

FIGURE 5.3.9 Hexagonal prism with the side-faces all squares.

Prisms could also have all of their side-faces made of congruent equilateral triangles. Such prisms are sometimes called antiprisms.

Exercise: Draw (sketch) an antiprism with regular squares at the top and the bottom and with side-faces all made of congruent equilateral triangles.

If we exclude the prisms and antiprisms, then the number of semiregular polyhedra decreases drastically. Discounting the Platonic solids (which are certainly semiregular) and some *chirals* (to be defined in the next paragraph), there are 13 semiregular polyhedra that are not prisms. These regular polyhedra are sometimes called **Archimedean solids**. We exhibit them all, along with their long names in Figure 5.3.10.

The three-dimensional mirror images of 11 of the above 13 solids are congruent to the originals. The two exceptions are the snub cube and the snub dodecahedron. Solids that are not congruent to their mirror images are sometimes called *chirals*. The two chirals that are the mirror images of the snub cube and the snub dodecahedron complete the list of semiregular polyhedra.

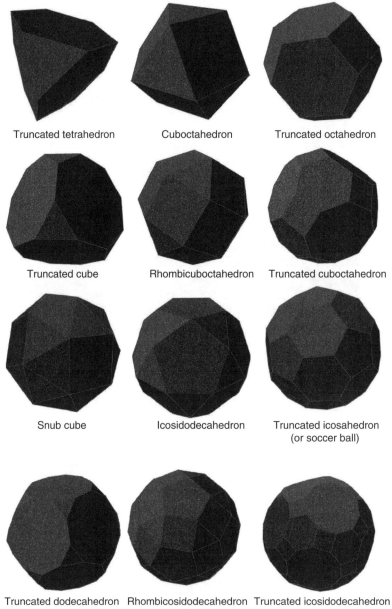

Truncated tetrahedron   Cuboctahedron   Truncated octahedron

Truncated cube   Rhombicuboctahedron   Truncated cuboctahedron

Snub cube   Icosidodecahedron   Truncated icosahedron
(or soccer ball)

Truncated dodecahedron   Rhombicosidodecahedron   Truncated icosidodecahedron

Snub dodecahedron

FIGURE 5.3.10 Semiregular polyhedra (or Archimedean solids).   □

A List of Platonic and Archimedean Solids

In the following table, we list all the Platonic and Archimedean solids:

Solid	Types of Faces	Number of Faces	Number of Edges	Number of Vertices
Tetrahedron	4 triangles	4	6	4
Cube	6 squares	6	12	8
Octahedron	8 triangles	8	12	6
Dodecahedron	12 pentagons	12	30	20
Icosahedron	20 triangles	20	30	12
Truncated tetrahedron	4 triangles and 4 hexagons	8	18	12
Cuboctahedron	8 triangles and 6 squares	14	24	12
Truncated octahedron	6 squares and 8 hexagons	14	36	24
Truncated cube	8 triangles and 6 octagons	14	36	24
Rhombicuboctahedron	8 triangles and 18 squares	26	48	24
Truncated cuboctahedron	12 squares, 8 hexagons, and 6 octagons	26	72	48
Snub cube	32 triangles and 6 squares	38	60	24
Icosidodecahedron	20 triangles and 12 pentagons	32	60	30
Truncated icosahedron	12 pentagons and 20 hexagons	32	90	60
Truncated dodecahedron	20 triangles and 12 decagons (ten-sided polygons)	32	90	60
Rhombicosidodecahedron	20 triangles, 30 squares, and 12 pentagons	62	120	60
Truncated icosidodecahedron	30 squares, 20 hexagons, and 12 decagons	62	180	120
Snub dodecahedron	80 triangles and 12 pentagons	92	150	60

Euler Characteristic

Consider a simple cube. It has 6 faces, 12 edges, and 8 vertices. At this moment it would appear that there is no clear reason why we are interested in the following algebraic expression.

$$\text{(the number of faces)} - \text{(the number of edges)} + \text{(the number of vertices)}$$

In the cube we have $6 - 12 + 8 = 2$. Now, take a look at, say, the octahedron: this time we find 8 faces, 12 edges, and 6 vertices. Computing again (the number of faces) − (the number of edges) + (the number of vertices) we get $8 - 12 + 6 = 2$, the same number as above. Is this a coincidence? Let us postpone the answer to this question for a couple of paragraphs.

If F is the number of faces, E the number of edges, and V the number of vertices in a polyhedron, then the number $F - E + V$ is called the ***Euler characteristic*** of the polyhedron.

Simple computations using the table we provided above show that the Euler characteristic of the surface of every Platonic or Archimedean solid is 2. However, it is not true that the Euler characteristic of the surface of every nonconvex polyhedron is 2, and a counter-example is provided in Exercise 4 by the end of this section.

Clearly, the Euler characteristic is a kind of measure of surfaces that have polygonal faces. What does it measure? We will take another look at this problem in Sections 6.2 and 6.3 and provide some loose answers. At this point, we note that the Euler characteristic has nothing to do with the geometry of the objects (e.g., with the flatness of the bounding polygons)—it has a lot to do with their *topology* (the internal structure of the points within the objects).

A Bit of Math. The Invariance of the Euler Characteristic
The main reason behind the invariance of the Euler characteristic of, say, convex polyhedra, is the following proposition that we will not prove.

Proposition: Every convex polyhedron can be changed to any other convex polyhedron by means of the moves described in Figure 5.3.11, combined with extending, shortening, protruding or flattening of edges or faces

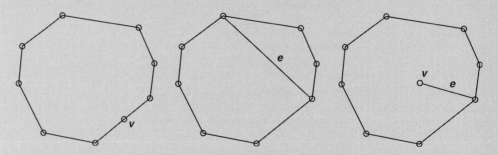

FIGURE 5.3.11. **1.** Subdivide an edge into two edges by inserting a vertex in its interior (leftmost figure) or the reverse of this procedure. **2.** Subdivide a face into two by joining two vertices (middle figure) or the reverse of this procedure. **3.** Insert a vertex in the interior of a face and an edge and join that vertex with another vertex in that face (rightmost figure) or the reverse of this procedure.

It is very easy to show that each of the procedures 1-3 described in the caption of Figure 5.3.11 does not change Euler characteristic. It is then an immediate consequence of the above proposition that all convex polyhedra have the same Euler characteristic.

□

Exercise: Use the moves described in the caption of Figure 5.3.11 to change a cube into a tetrahedron. [*Hint.* First change the cube into a pyramid with a triangular basis.]

The illustrations in Figures 5.3.12 and 5.3.13 exhibit apparent influence of the geometry of Archimedean solids.

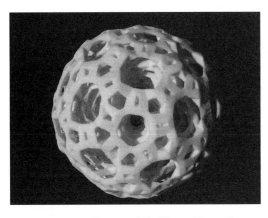

FIGURE 5.3.12 George W. Hart, *Deep Structure*, 2002.

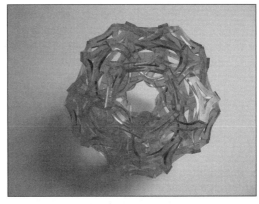

FIGURE 5.3.13 George W. Hart, *Cagework 1*, 2005.

Exercises:

1. How high is a pyramid over a square base if all its edges are equal to 1 m? What is its volume?

2. a. Subdivide a cube into two congruent parts by cutting it with a plane that does not pass through any of its eight vertices.

   b. Subdivide an icosahedron into two congruent parts by cutting it with a plane that does not pass through any of its 14 vertices.

   c. Subdivide a tetrahedron into two congruent parts by cutting it with a plane that does not pass through any of its four vertices.

3. The edges of the "dual polyhedron" of a Platonic solid can be obtained by joining the centers of the adjacent faces of the Platonic solid (in case of an isosceles triangle the center is the intersection of the three heights). Find the duals to all five Platonic solids.

4. Find the Euler characteristic of the surface of the following (three-dimensional) object made of cubes (Figure 5.3.14).

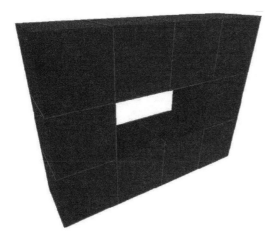

FIGURE 5.3.14

5. What is the Euler characteristic of the object we get by (miraculously) drilling out a smaller cubic shape from the middle of a cube?

6*. You have an unlimited number of cubic boxes of the same size, which you use as bricks to build a structure. Initially you start with one cubic box and then you glue other boxes one by one, so that at every stage any two touching cubes share either a full face, or a full edge or a single vertex. Faces of boxes are glued together with their edges and vertices—we never glue a vertex or an edge that does not belong to a face that is to be glued. One face of a new box is glued to one face of one of the already used boxes, or two or more faces of the new box are glued to the corresponding number of faces of the used boxes. In the latter case, we stipulate that the faces in the used boxes that are being glued make a connected (one-piece) structure. Show that whatever structure you construct following the above rules, its Euler characteristic must be equal to two.

7. a. Sketch a tiling of the three-dimensional space with all tiles being copies of a prism with an isosceles triangle as a base and with squares as side-faces.

   b. Do the same as in (a), but this time use copies of a right hexagonal prism (as shown in Figure 5.3.9) as tiles.

5.4 SPHERE, CYLINDER, CONE, AND CONIC SECTIONS

In this section, we take a cursory look at some objects in three-dimensional space. We will revisit and expand this subject later (in Chapter 6).

Sphere

A **sphere** of radius r is the set of all points in the (three-dimensional) space at distance r from a fixed point in the space. A sphere is a surface, not a solid object. A related object is a ball: a **ball** of radius r is the set of all points in the space that are at distances at most r from a fixed point in the space (Figures 5.4.1 through 5.4.4).

FIGURE 5.4.1 Here is a sphere ….

FIGURE 5.4.2 … and here is a ball. We cannot see any difference ….

FIGURE 5.4.3 … until we cut them open: the sphere to the left and the ball to the right.

In Figure 5.4.4 we show two views of an artwork done on a spherical canvas.

FIGURE 5.4.4 **(See color insert following page 144.)** (Art on a spherical canvas) Dick Termes, *Emptiness*, 24″ sphere, 1986.

In the rest of this section, we only consider surfaces.

Cylinder

Here is a picture of an open cylinder (a cylinder without base or top; Figure 5.4.5).

It is easy to get a cylinder from a rectangular piece of flexible material by gluing a pair of opposite edges as shown in the sequence of pictures in Figure 5.4.6. We will use this technique of gluing edges a few more times in Chapter 6.

FIGURE 5.4.5

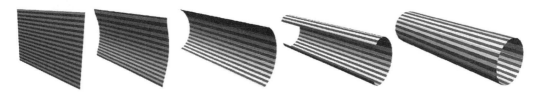

FIGURE 5.4.6

Cone

In Figure 5.4.7, we show a part of an open cone ("open" in the sense that there is no base). The cones we consider here have surfaces that extend unboundedly, becoming wider and wider.

Cones are important objects from a historical point of view. During the pre-Descartes times (seventeenth century and earlier) they were used as the main route to study certain important planar (two-dimensional) curves: circles, ellipses, parabolas, and hyperbolas. All of these curves can be obtained by intersecting one or two cones (details below) with suitably chosen planes—so, they are called ***conic sections***. For example, in Figure 5.4.7, we see (or suspect) that we get circles if we cut the cone with horizontal planes.

FIGURE 5.4.7

After the invention of coordinate systems (Descartes, around 1625), it turned out that conic sections in some way exhaust all possible curves definable analytically by certain equations (the so-called *quadratic* curves in the plane; more about that is given further in this section).

We will introduce a bit of terminology and conventions before we exhibit all of the conic sections. The sharp top of a cone is called an ***apex***. A semiline starting at the apex of a cone and lying on its surface is called the ***slant height*** of the cone. The semiline starting at the apex and passing through the centers of the circular cross sections is called the ***axis*** of the cone. If the axis is perpendicular to the base of the cone, then we will say that the cone is ***right***. Our cones will always be right cones unless otherwise stated. The object made of two cones attached by their apexes and sharing the same axis line is called a ***double cone***. Each cone in a double cone is called a ***nappe***. □

Conic Sections

We will now cut cones and double cones with planes to get conic sections. Our conic sections will be nondegenerates, which means that we will not cut with planes that intersect the cone or the double cone in two lines, a line, or a point. Details are given in Figures 5.4.8 through 5.4.11 and their captions.

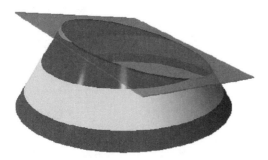

FIGURE 5.4.8 If the plane intersects all the slant heights of a cone, then the crosscut is an ellipse.

FIGURE 5.4.9 If we cut a (right) cone with a plane that is perpendicular to its axis, then the resulting cross section is a circle. This is a special case of an elliptical cross section.

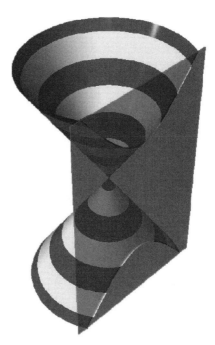

FIGURE 5.4.10 In order to get a parabola as a conic section, we need to choose our plane carefully. First, the plane in the background (to the left) is a tangent plane: it touches the cone along a slant height. Now, in order to get a parabolic cross section we need to choose a plane that intersects the cone and is parallel to some tangent plane. The two planes we see above are parallel (remember that all the three-dimensional objects are represented using perspective).

FIGURE 5.4.11 Crosscutting a double cone with any plane that intersects both of the nappes yields a hyperbola. Keep in mind that the cones extend unboundedly—so parabolas and hyperbolas are unbounded curves too.   ☐

A Bit of Math. Regarding Algebraic Descriptions of Conic Sections
We assume here that the reader is familiar with (planar, rectangular) coordinate systems (a brief introduction is given in Section 2.2). A circle of radius r can be set in a coordinate system in such a way that every point (with coordinates) (x, y) on the circle satisfies the equation $x^2 + y^2 = r^2$. Every ellipse can be put in a coordinate system such that its equation is $\frac{x^2}{a^2} + \frac{y^2}{b^2} = 1$ (a and b are constants). Parabolas can be defined by $y = cx^2$ (for some constant c), whereas hyperbolas by $xy = d$ for some constant d. It turns out that every polynomial equation in x and y of second degree (meaning that the highest exponent involving x or y is 2) either defines one of the four conic sections described above, or degenerates into an equation defining a line or two lines (example $y^2 = x^2$), a point (example $y^2 + x^2 = 0$), or no points at all (example $y^2 + x^2 = -1$). So, excluding the degenerate cases, conic sections are all of the curves definable by polynomial equations of second degree. For example, both $2x^2 + 3xy - y^2 - 2y + 1 = 0$ and $x^2 + xy - x = 0$ must define some conic sections. ☐

Using Tangents to Draw Conic Sections. Circle and Ellipse

A. Circle. Start with a square (see Figure 5.4.12). Extend the lines through the vertical edge of the square and mark points on them as follows: the middle point of each of the two vertical edges of the square is marked by 0, and the top vertices by 1; then any other point on this line is marked by the number that measures its distance to 0 using one half of the (vertical) edges as one unit of length, with positive numbers used above 0, and negative numbers used below 0. The initial square is visible in Figure 5.4.12. The rest is easy: simply join any point marked by a number m on the right-hand vertical line with the point marked by $1/m$ on the left-hand vertical line.

The resulting line segments will be tangent to the circle of radius one unit long, and centered at the center of the square. The more line segments we draw the clearer the outline of the circle will be. In Figure 5.4.12 you see that, for example, the point on the right-hand vertical line marked by the number 2/8 is joined to the point marked by the reciprocal 8/2; similarly, the point on the right-hand line marked by 7/8 is connected to the point on the left-hand vertical line marked by 8/7. Note that the same procedure is used for the bottom part of the circle—only the numbers we use there are negative.

B. Ellipse. We can easily modify this method to draw ellipses. Replace the starting square with a rectangle and repeat the above procedure. A sketch is given in Figure 5.4.13.

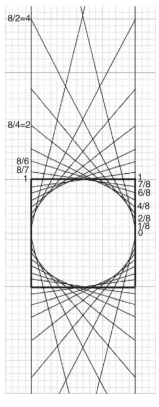

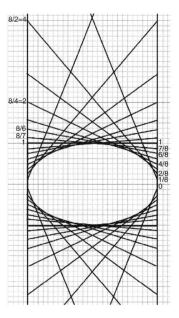

FIGURE 5.4.12   FIGURE 5.4.13

A Bit of Math. Justifying the Construction of the Circle

We will now show that the lines in the construction A are indeed tangential to a circle of radius 1. To that end, first set the circle in a coordinate system, so that its center is at the origin. As we have seen, the equation of such a circle of unit radius is $x^2 + y^2 = 1$. The two vertical lines that touch the circle have equations $x = -1$ and $x = 1$, respectively. It suffices to show that every nonvertical tangent t to the circle intersects these two lines at points C and D (respectively) with mutually reciprocal second coordinates (see Figure 5.4.14: using the notation from this picture, we need to show that the distances from C and D to the x-axis are indeed mutually reciprocal numbers m and $\frac{1}{m}$ as indicated in the picture). The proof we provide below is rather straightforward and the reader should not be discouraged by its length or by the seemingly complicated formulas that we encounter.

Start by choosing a line l passing through the origin and intersecting the circle at a point B (as in the picture). If the line l is vertical, then the tangent line t is horizontal and both C and D are at a distance of 1 unit from the x-axis and so these two numbers (both equal to 1) are mutually reciprocal. So, we can assume that l is not vertical. Any line l that is not vertical and that passes through the origin has equation $y = kx$, where k is the slope of the line. We may assume that k is positive; the other case is symmetric and needs only a minor modification of the following argument. So, the intersection point B is in the first quadrant as shown in Figure 5.4.14 (so that both coordinates of B are positive). The point B, being a point on the line l, satisfies the equation $y = kx$. However, since B is also on the circle it satisfies $x^2 + y^2 = 1$. Solving these two equations simultaneously is easy, and the solution is $x = \dfrac{1}{\sqrt{1 + k^2}}$ and $y = \dfrac{k}{\sqrt{1 + k^2}}$. So, B has coordinates $\left(\dfrac{1}{\sqrt{1 + k^2}}, \dfrac{k}{\sqrt{1 + k^2}} \right)$. Now we pay attention to the tangent line t. Since it is perpendicular to l, it has a slope of $-\dfrac{1}{k}$. We use the following fact from basic geometry (or linear algebra): a line with a slope s and passing through a point (a, b) has equation $\dfrac{y - b}{x - a} = s$. In our setting, since t has a slope of $-\dfrac{1}{k}$ and passes through the point $\left(\dfrac{1}{\sqrt{1 + k^2}}, \dfrac{k}{\sqrt{1 + k^2}} \right)$, we get the following equation of the line t: $\dfrac{y - \dfrac{k}{\sqrt{1 + k^2}}}{x - \dfrac{1}{\sqrt{1 + k^2}}} = -\dfrac{1}{k}$.

In order to find the intersection point C, we solve the system given by the last equation for t and the equation $x = 1$ of the vertical line to the right of the circle. This is also straightforward (we simply substitute $x = 1$ in the equation for t), and the solution (the second coordinate of C) is $\dfrac{-1 + \sqrt{1 + k^2}}{k}$. A very similar argument for the point D (where we should use $x = -1$ instead of $x = 1$) gives $\dfrac{1 + \sqrt{1 + k^2}}{k}$ for the second coordinate of D. Multiplying $\dfrac{-1 + \sqrt{1 + k^2}}{k}$ $\dfrac{1 + \sqrt{1 + k^2}}{k}$ and simplifying yields 1 as the result, so that these two numbers are indeed mutually reciprocal as claimed. $\square$

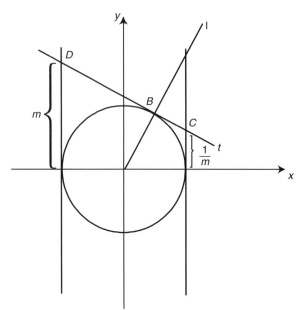

FIGURE 5.4.14

Using Tangents to Draw Conic Sections. Parabola

This conic section is the easiest to sketch using tangents (Figures 5.4.15 and 5.4.16). We need a couple of line segments of equal length and that share a common end vertex. Subdivide both of the line segments into an equal number of smaller segments of equal length. In Figure 5.4.15, both of the two initial segments are subdivided into eight segments of equal length. The larger the number of segments in the subdivisions, the better the approximation of the parabola with tangents, and the smoother the appearance of the curve with tangents. Number the points of subdivision: for one of the two segments start from the common endpoint, for the other start from the other endpoint (see Figure 5.4.15). Finally, join each point marked by a number on one of the two starting segments with the point on the other starting segment labeled by the same number.

In Figure 5.4.16 we use tangent to construct two parabolas and one circle.

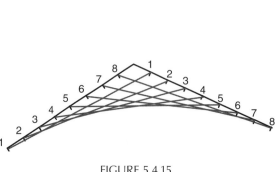

FIGURE 5.4.15

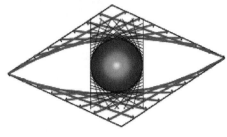

FIGURE 5.4.16 Two parabolas and a circle, all done using tangents as explained above. □

Hyperbola. A Construction

The construction described below and illustrated in Figure 5.4.17 is of a different type compared to the preceding three constructions: we will not use tangents. In this case we will have the precise position of some points on the hyperbola that we construct.

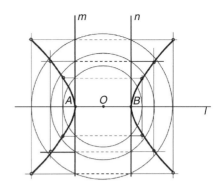

FIGURE 5.4.17

We start with a pair of parallel lines (*m* and *n* in the picture), and one line that is perpendicular to both (the line *l* in the picture). The intersection points *A* and *B* are already points of the hyperbola we are about to construct. Identify the middle point *O* of the segment *AB*. Choose any circle centered at *O* and intersecting *m* and *n*. Draw horizontal lines through the four points of intersection of the circle with *m* and *n*, and draw vertical lines through the points of intersection of the circle with the line *l*. These four lines will have four intersection points, and these intersection points all lie on a hyperbola. Repeat the procedure with another circle to get four more points. Performing this repeatedly will give us more points on the hyperbola (in the illustration the points we have constructed are indicated with small circles). Then join them with a smooth curve as is done in Figure 5.4.17. □

In Figure 5.4.18, line segments are used to generate curves in three dimensions.

FIGURE 5.4.18 Charles Perry, _*Solstice*, Barnet Plaza, Tampa, FL, 28 Feet, 1985.

Cardioid. Two Constructions

We give two constructions of a heart-like curve called a **cardioid**: one with tangents (Figure 5.4.19), the other with circles (Figure 5.4.20).

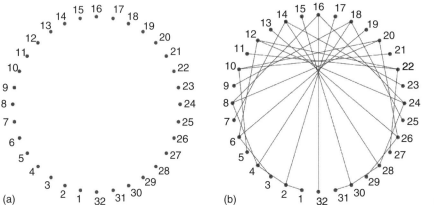

(a)   (b)   FIGURE 5.4.19

A. Using tangents. We start with a circle and choose a number of equally spaced points on the circle (and we do not need the circle anymore). In Figure 5.4.19a, there are 32 chosen points. Label the points with consecutive numbers starting with 1. Now join the point numbered 1 with the point numbered 2, the point numbered 2 with the point numbered 4, the point numbered 3 with the point numbered 6, and keep going (see Figure 5.4.19b). Eventually, we join the point 16 with the point 32. We continue: following the recipe, the point 17 should be joined with the point 34, except that there is no point 34. We join 17 with the point labeled by the remainder we get after we divide $(2)(17) = 34$ by 32 (that is, with 2). Now we continue like that: 18 is to be joined with the remainder we get after we divide $(2)(18) = 36$ with 32, which is 4. So, 18 should be joined with 4. Continue until we exhaust all of the numbers we use to label points. The contour of a cardioid is visible in Figure 5.4.19b. The picture in Figure 5.4.20a is obtained by applying the same procedure starting with 256 points (instead of 32 as above).

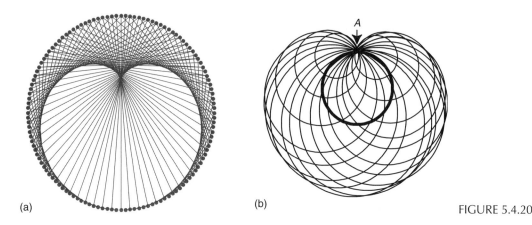

(a)   (b)   FIGURE 5.4.20

B. Using circles. Draw any circle and fix a point A on that circle (the thick circle in Figure 5.4.20b). Then draw circles centered at various points on the starting circle and passing through the point A. That is all! Figure 5.4.20b is obtained by applying this method.   □

Exercises:
1. How should we cut a double cone with one plane to get only one intersection point? How should we cut it to get one intersection line? What about two lines?
2. Identify all cylindrical sections (intersections of an unbounded cylinder with planes), and explain how to get them.
3. Use tangents (as in Figure 5.4.13) to construct an ellipse inscribed within a golden rectangle.
4. Using the methods described in this section, draw the lower left quarter of an ellipse in the lower left rectangle, a parabola in the lower right rectangle, and an upper half of a circle in the upper rectangle to get a continuous curve. (You may want to copy Figure 5.4.21 on a large piece of paper.)

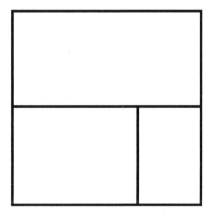

FIGURE 5.4.21

5. Use a ruler and a compass to construct 16 equidistant points on a circle, and then use these points to draw an outline of a cardioid.
6. Using the technique of constructing a parabola as an envelope of tangents, draw a parabolic antenna with the baselines as shown in Figure 5.4.22. (You may add the base support by freehand.)

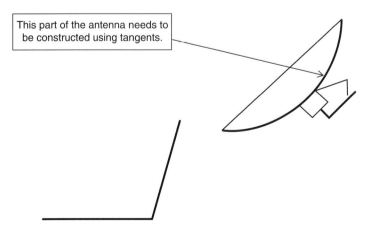

This part of the antenna needs to be constructed using tangents.

FIGURE 5.4.22

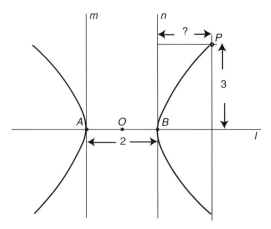

FIGURE 5.4.23

7. Consider the construction of a hyperbola outlined in Figure 5.4.17, and suppose that the distance between the points *A* and *B* is 2 units (see Figure 5.4.23). Suppose that a point *P* on the hyperbola is 3 units above the horizontal line *l*. How far is *P* to the right of the line *n*?

Extra: A Construction with Circles. How to Draw an Egg Or an Oval

In the figure to the below, we show a method for drawing an egg-like shape using just circles. We start with the two large circles, and then construct two smaller circles as indicated. The arcs we use in the oval are in full stroke, and the rest of the circles are in dashed stroke.

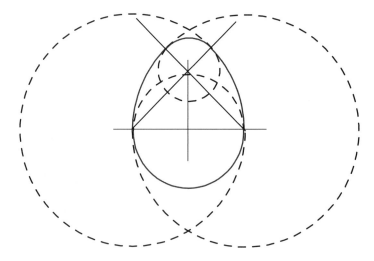

5.5 GEOMETRY, TILINGS, FRACTALS, AND CELLULAR AUTOMATA IN THREE DIMENSIONS

In the first four chapters, we were almost exclusively concerned with planar objects. We covered planar Euclidean geometry, and we saw a two-dimensional model of the hyperbolic geometry. Subsequently we listed various planar tilings, both of the Euclidean plane and of the hyperbolic plane. Our fractals were also mostly planar, and our cellular automata did not go past the second dimension. There are a number of good reasons why we did not venture into higher dimensions. For example, planar objects are more accessible and easier to depict. Moreover, notions and ideas are generally of increasing complexity as we move into higher dimensions. In this section, we will give a brief, mostly visual overview of some of the three-dimensional analogs of a few of two-dimensional objects that we covered.

Euclidean Geometry, Hyperbolic Geometry, and Inversion in Three Dimensions

The axioms of Euclidean and hyperbolic geometry in three dimensions make use of one more primitive notion (besides the notions of *point* and *line*): that of a plane. The difference between the two geometries is in the three-dimensional analog of the fifth Euclidean axiom, stating that given a plane and a point outside that plane, there is exactly one plane containing the given point and parallel to the given plane. As one would suspect, this claim is true in the Euclidean geometry, whereas it fails in the hyperbolic geometry, where there are infinitely many hyperbolic planes passing through a given point and parallel to a given hyperbolic plane. Predictably, in some models of three-dimensional hyperbolic geometry, hyperbolic planes are parts of spherical surfaces. For example, a view of a regular hyperbolic dodecahedron is shown in Figures 5.5.1 and 5.5.2.

FIGURE 5.5.1 A hyperbolic dodecahedron. Compare with the regular dodecahedron (Figure 5.3.5) as well as with the equilateral hyperbolic triangle (Figure 4.4.5).

FIGURE 5.5.2 A wild, thorny hyperbolic dodecahedron.

Inversion of points in space with respect to a sphere can be defined similarly as the two-dimensional inversion with respect to a circle, and its role in the three-dimensional Poincaré model is similar to the role of the circle inversion with respect to the two-dimensional Poincaré model we covered in Chapter 4. We illustrate this concept in Figures 5.5.3 and 5.5.4.

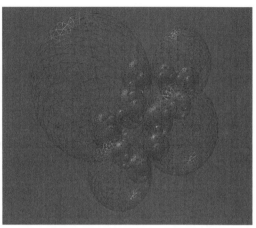

FIGURE 5.5.3 (**See color insert following page 144.**) Jos Leys, *Balinv126,* 2005. This is an approximation of the image of a set of spheres under inversion with respect to a sphere.

FIGURE 5.5.4 Here is a three-dimensional analog of Figure 4.2.16: in this case we apply inversions with respect to spheres.

Space Tilings

Symmetries in three dimensions are also defined as being transformations preserving distances—only this time they move points in the three-dimensional space. The same treatment as in the planar case leads to the concept of groups of symmetries of three-dimensional objects as a measure of how *symmetric* the objects are.

We move briefly to the subject of tilings of the three-dimensional space. The role of polygons is now taken by filled polyhedra. In the three-dimensional case, we stipulate that every point in the space is within some polyhedron and that two adjacent polyhedra can only share a common vertex, a common edge, or a common face. It is relatively easy to see that we can fill the three-dimensional space with congruent cubes (Figure 5.5.5). The cube

FIGURE 5.5.5 A part of the tiling of three-dimensional space with congruent cubes.

is the only Platonic solid that tiles (nicely packs) three-dimensional space. Congruent tetrahedrons do not tile (or pack) three-dimensional space. It is interesting to note that Aristotle, in his treatise *On the Heavens*, claimed otherwise.

Moving to Archimedean solids, there is again only one of them that tiles the three-dimensional space: the truncated octahedron. We show a part of that tiling in Figure 5.5.6.

In the above two tilings of the three-dimensional space, we have used tiles of a single type (shape and size). There are other tilings of the three-dimensional space with all tiles being copies of a single polyhedron. In Figures 5.5.7 and 5.5.8 (and Animation 5.5.1), we show a tiling of the three-dimensional space with a stellated rhombic dodecahedron.

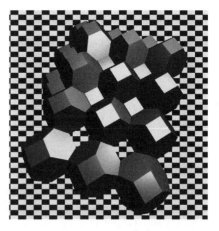

FIGURE 5.5.6 The truncated octahedron tiles the three-dimensional space.

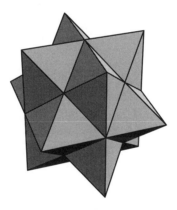

FIGURE 5.5.7 A stellated rhombic dodecahedron.

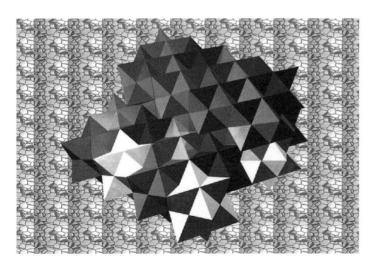

FIGURE 5.5.8 AND ANIMATION 5.5.1 A tiling of the three-dimensional space with a stellated rhombic dodecahedron.

Fractals and Cellular Automata in Three Dimensions

Fractals in three dimensions are again defined through the property of being self-similar, where in this case a *similarity* is an angle-preserving transformation of the points of the three-dimensional space. The concept of cellular automata also has a straightforward generalization in three dimensions, only this time we start with a planar grid of white and black cubes and a rule that determines how the colors of the next planar configuration of cubes depend on the colors of the cubes in the preceding planar configuration. As was the case in two dimensions, in the three-dimensional game of life the cellular automation rules depend on the number of adjacent cubes in each configuration. We give examples of three-dimensional fractals in Figures 5.5.9 through 5.5.12, and an example of a three-dimensional cellular automation, illustrating a three-dimensional game of life, in Figures 5.5.13 through 5.5.15.

FIGURE 5.5.9 A Menger sponge: we show the third iteration of the drilling procedure. The fractal results after infinitely many iterations.

FIGURE 5.5.10 Jean-Charles Marteau, *Menger Spheres*, 2003.

FIGURE 5.5.11 The three-dimensional version of the Sierpinski triangle; four iterations.

FIGURE 5.5.12 **(See color insert following page 144.)** Marcus Vogt, *Tower Series 04*, XenoDream 3D fractal software, then Photoshop, 2007.

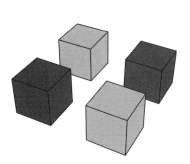

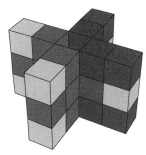

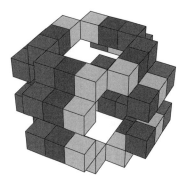

FIGURE 5.5.13 **(See color insert following page 144.)** In the initial configuration we start with four colored cubes; we do not show the white cubes.

FIGURE 5.5.14 **(See color insert following page 144.)** A cube becomes or stays colored if it is adjacent to exactly two colored cubes. This is what we get in the next step.

FIGURE 5.5.15 **(See color insert following page 144.)** One more step: the population of the colored cubes spreads.

Exercises:

1. Draw precisely in a two-point perspective, a Menger sponge after the first iteration. [*Hint*: see the construction of a square in Section 5.1 as well as Exercise 8 in the same section.]
2. Describe a tiling of the three-dimensional space with the tile shown in Figure 5.5.16.
3. In Figure 5.5.17, we depict a right pyramid over a square base. The length of the side of the base is 1 unit and the height of the pyramid is 0.5 units. Describe a tiling of the three-dimensional spaces with the pyramid.

FIGURE 5.5.16 The tile is made of four cubes, one of which (in the corner) is not visible.

FIGURE 5.5.17 A right pyramid over a square base.

4. Describe a tiling of the three-dimensional space with polyhedra, such that the tiling is at the same time a complete fractal. [*Hint*: you need infinitely many types of tiles for such a tiling.]
5. Start with a cluster of 4 × 4 cubic cells, and consider the cellular automation defined by the following rule: a cubic cell survives or is born only if it has exactly two neighbors (where neighboring cubes share a face, an edge, or a vertex). Describe the positions of the living cubic cells after one iteration.

CHAPTER 6

Topology

Topology is a study of internal structures of objects and mutual position of the points within objects. While in geometry it is important how far a point is from another point, or if the object is *straight* or *curved*, in topology that is almost irrelevant. For example, from a topological point of view, a straight line is the same as a bent line: the mutual, internal relationships of the points in both objects are essentially the same. Similarly, in topology, a sphere and a crumpled sphere can be considered as being of the same internal structure. In this chapter, we will provide intuitive definitions of some main topological notions, and then we will briefly and informally consider various topological spaces and some of their properties. The reader should be aware that all the notions we encounter in this chapter have formal and precise counterparts within *topology* and *algebraic topology*, the two important mathematical areas.

It is common in topology to refer to objects as (topological) spaces. We will use both terms (objects and spaces) interchangeably. We will continue to refer to the *three-dimensional space* relying on our intuitive notion of it, and without properly defining it.

6.1   HOMOTOPY OF SPACES: AN INFORMAL INTRODUCTION

The Notion of Homotopy

Consider the letters A and R as planar objects made of (mathematical) points. As we see in Figure 6.1.1, these two letters can be *deformed* one into the other.

AAAAARRRRR

FIGURE 6.1.1 A few steps in continuous deformation from A to R.

Since A can be deformed continuously to R, we say that these two objects (or spaces) are homotopic. Informally, two spaces are **homotopic** if we can continuously deform one of them into the other without cutting or tearing and without pasting. The process

FIGURE 6.1.2 We may stretch the point at the left into a (curved) line segment, and we may continue stretching, shrinking, and bending the line segment. However …

FIGURE 6.1.3 … we are not allowed to glue or patch, as it is apparently done here.

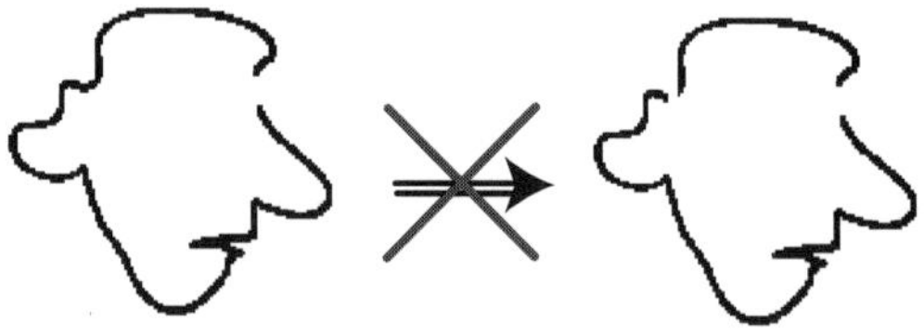

FIGURE 6.1.4 … nor are we allowed to cut the line segment, as is done in this illustration.

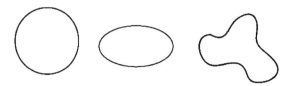

FIGURE 6.1.5 We may deform a circle into an ellipse, and we may bend it. All the spaces we get in such a way will be homotopic.

FIGURE 6.1.6 However, a circle, be it very small, is not homotopic to a point. The last step of continuously deforming a circle into a point involves a kind of pasting, which we also exclude.

of deforming one space into another without cutting or pasting is called a **homotopy**. In Figures 6.1.2 through 6.1.8, we explain visually what we mean by **deforming** a space into another, and, in particular, what we mean by the stipulation that cutting or pasting is not allowed.

As indicated in Figure 6.1.6, a circle is not homotopic to a point. However, any disk (the circle together with its interior) is homotopic to a point: there is no cutting or gluing involved in shrinking a disk into a point.

In Figure 6.1.8, we see why an open cylinder is homotopic to a circle.

FIGURE 6.1.7 We may shrink, enlarge, or deform a sphere in any way. As long as we do not cut or glue, we always get a space homotopic to the starting sphere. Making a hole as in the figure to the right is considered cutting, and shrinking a sphere to a point is considered pasting. So, the sphere to the left is homotopic to the middle one, and none of them is homotopic to the punctured sphere to the right.

FIGURE 6.1.8 A homotopy between a cylinder and a circle: no cutting, no pasting.

We move on to the notion of homotopy *within* a larger space. Consider the example in Figure 6.1.9 (and its caption).

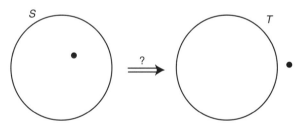

FIGURE 6.1.9 Is the space S to the left, made of a point in a circle, homotopic to the space T to the right where the point is out of the circle? □

The answer to the question stated in the caption of Figure 6.1.9 is affirmative: take the point out of the plane containing the circle and simply lift it above the circle, and then drop it back into the plane and out of the circle. The two-piece space S to the left is thus homotopic to the two-piece space T to the right, with the homotopy performed in three dimensions. Could we do the same in two dimensions? More precisely, could we deform S to T while being confined within the plane containing S and T all the time? Obviously* not! The only

---

* As every mathematician knows, "obvious" is perhaps the most dangerous word. The claims we make here have precisely stated counterparts, the statements (and the justifications) of which are well beyond our scope. Indeed, the statement that the two two-piece objects in Figure 6.1.9 are not homotopic within the plane is something that deserves a formal justification.

way to take the point out of the circle without leaving the plane, and without cutting the circle, is first to merge it with the circle (deformed or not) and then tearing it off the circle and out of it. That would not be a homotopy since it would involve pasting and cutting.

The example in Figure 6.1.9 and the commentary illustrates a new notion: that of homotopy within a larger space. Two spaces *A* and *B* lying in a larger space *X* are said to be **homotopic within** *X* if there is homotopy from *A* to *B* such that all the intermediate spaces during deformation stay within the space *X**.

Most of the time the underlying space *X* will be either within our drawing/writing plane or in the three-dimensional space we perceive. We will continue to refer to the latter as simply the *three-dimensional space*.

Comparing the concept of *homotopy* with the concept of *homotopy within a larger space*, we note that the former is more general: if two spaces are homotopic within a fixed larger space, then they are certainly homotopic. However, the converse is not true. One may say that two spaces are homotopic if we can continuously deform (no cutting or pasting) the inner structure of one into the other, without necessarily setting the two spaces within a larger space where that process is done. It is clear that the concept of *homotopy within a larger space* is more intuitively accessible, and so we will deal with that one most of the time.

As we saw above, the space consisting of a circle and a point within it is homotopic within the three-dimensional space to the space consisting of a circle and a point out of the circle, but these two are not homotopic within the plane containing them. We will now consider the notion of *homotopy within a space* in a few more examples.

Example 1

Consider the spaces *L* and *R* depicted in Figure 6.1.10.

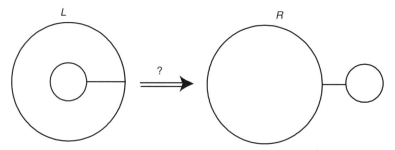

FIGURE 6.1.10 Is the space *L* homotopic to the space *R* within the three-dimensional space? Are the spaces homotopic within the plane that contains both of them?

It is not hard to see that the answer to the first question in the caption of Figure 6.1.10 is affirmative. To deform *L* to *R* it suffices to rotate (in three dimensions) the inner circle of *L* together with the horizontal line segment through 180° around the vertical tangent line passing through the point where the line segment touches the larger circle. ☐

---

* Our notion of "homotopy within *X*" is a cross between the topological concepts of homotopy (described here) and ambient isotopy (not described here).

The second problem, where we ask for homotopy *within* the drawing plane, is more difficult. Perhaps surprisingly, such a homotopy exists! We leave it as an exercise; a hint will be provided in the next example.

Example 2

In Figure 6.1.11 (and Animation 6.1.1), we see two entangled rings (hollow surfaces only), joined by a tubular link. In the next illustration (Figure 6.1.12), we see a space in which these two rings are untangled. Is the space L depicted in Figure 6.1.11, homotopic to the space R depicted in Figure 6.1.12? The answer to that question is an emphatic "yes": there is such a homotopy in four dimensions! In order to see what we mean by the last statement, consider again the spaces depicted in Figure 6.1.9. We noted in that example that taking the point out of the circle can be easily done in three dimensions, but there is no such homotopy in two dimensions. Now, imagine for a moment that you are a two-dimensional creature, existing within the drawing plane. It would appear to such a simple creature that it is impossible to take the point out of the circle without cutting or gluing at some moment. That is, as we noticed above, true, if we restrict the homotopy to within the drawing plane. We are then faced with the following problem: how to explain to the poor two-dimensional creature that we could easily take the point out of the plane of the circle and put it back in the plane but out of the circle? The fictitious two-dimensional creature may not believe us, for the three-dimensional world would be out of his perceptional domain. Well, we are in the two-dimensional creature's shoes, so to speak: it is difficult for us to imagine that we can untangle the two rings depicted in Figure 6.1.11 by taking one of them out of the three-dimensional space into four dimensions, move it there a bit without any cutting or pasting, and then put it back in three dimensions so that the resulting space is as the one shown in Figure 6.1.12. Nevertheless, this is something that can be done mathematically, provided one has precisely defined models of the four-dimensional analog of the three-dimensional space. We will stay within the analogy we have provided, without giving technical details.

In fact, it is not necessary to believe that the space in Figure 6.1.11 is homotopic to the space in Figure 6.1.12 within four dimensions, for a homotopy exists in three dimensions! Such a homotopy is shown in Animation 6.1.1 (electronic version).

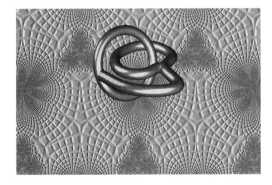

FIGURE 6.1.11 and ANIMATION 6.1.1 Can this space be untangled to the space shown in Figure 6.1.12 within three dimensions?

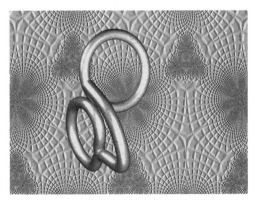

FIGURE 6.1.12 The rings are untangled in this space. ☐

Exercises:

1. Describe a homotopy that changes the space 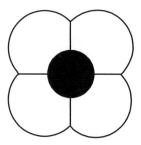 into the space **S**. Draw a few intermediate spaces as you deform the first space into the second.
2. Subdivide the set of all letters A–Z into homotopy classes. That is, decide which pairs of letters are homotopic, and which are not homotopic. If a pair of letters is homotopic, describe the homotopy (the deformation from one of the letters to the other) by drawing a few intermediate steps.
3. Show that the scissors in Figure 6.1.13 are homotopic within three dimensions to the eyeglasses frame (Figure 6.1.14) by drawing at least three in-between sketches showing how the scissors can be continuously deformed into the frame.

FIGURE 6.1.13 This is just an illustration of the usual, three-dimensional scissors.

FIGURE 6.1.14 The eyeglasses frame we refer to in Exercise 2 is also three-dimensional.

4. Show that the space shown in Figure 6.1.15 is homotopic to the space shown in Figure 6.1.16.

FIGURE 6.1.15 The space consists of all black points.

FIGURE 6.1.16 Only the black points belong to this space.

5. Figure 6.1.17 shows a design based on a Nazca geogliph in Peru (the original spiral is 250 ft. in diameter). Is it homotopic to a circle within the drawing plane?

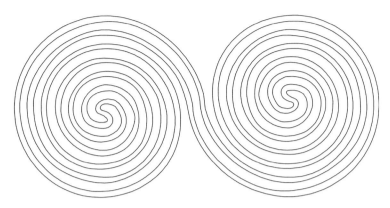

FIGURE 6.1.17 Two spirals.

6. Show that the space *L* shown in Figure 6.1.10 is homotopic *within the drawing plane* to the space *R* shown there. [*Hint*: see Figure 6.1.1 and apply similar homotopy, but within two dimensions.]

7. In Figure 6.1.18, we show a space within three dimensions consisting of two entangled circles connected with an arc. In Figure 6.1.19, the two circles are still connected with an arc but are not entangled anymore. Show that the two spaces are homotopic within three dimensions.

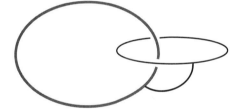

FIGURE 6.1.18 Two entangled circles connected with an arc.

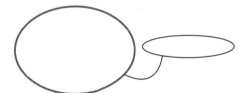

FIGURE 6.1.19 Two circles connected with an arc.

8*. The space in Figure 6.1.20 consists of three entangled circles connected with arcs. In the space depicted in Figure 6.1.21, the circles are not entangled. Show that the spaces are homotopic within three dimensions.

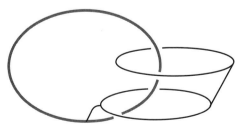

FIGURE 6.1.20 A space with three entangled circles connected with arcs.

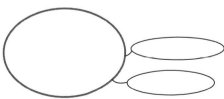

FIGURE 6.1.21 Three circles connected with arcs.

9. Figure 6.1.22 shows a punctured torus (the surface of a donut; only the surface!). Show that it is homotopic in three dimensions to the space made of the points in the numeral 8.

FIGURE 6.1.22 We cut a circular hole on the surface of a donut (surface only).

FIGURE 6.1.23 A knot.

10. Explain why the tubular knot shown in Figure 6.1.23 is homotopic within four dimensions to the torus. [*Hint*: read Examples 1 and 2 again.]

6.2 TWO-MANIFOLDS AND THE EULER CHARACTERISTIC

In the preceding section, we asked the reader to imagine being a two-dimensional creature living within the confines of a certain two-dimensional object. We will now use the same informal but intuitive approach to introduce the concept of a two-manifold.

Two-Manifolds

Consider a very large sphere, say, larger than the earth, and imagine we are miniscule, two-dimensional inhabitants of the surface of that sphere. There is no sky above us, and nothing below us. Being two-dimensional, our perceptional domain is just the surface of the sphere and nothing else. The surface of the sphere will then be our universe, and we may feel free to roam on that surface any way we want without ever realizing that our universe is but a small part of a larger, three-dimensional world (that in turn may be a part of a higher dimensional world). Owing to the relative sizes of the surface of the sphere and its imagined inhabitants, and because the sphere itself is positioned in three dimensions that we (as two-dimensional entities) cannot perceive, it would appear to us that our world is flat and that it extends unboundedly in all directions, precisely as if it were a plane. In other words, the surface of the sphere *feels locally* exactly the same as the surface of the plane. We arrived through a nonmathematical route to the following important intuitive definition.

A **two-manifold** is a space M that *feels locally* like the surface of the plane. Somewhat more precisely, that means that for every point A of M, the set of points closer to A than a fixed, small number of units, has *the same internal structure** as the space consisting of

---

* In the terminology of topology, any two spaces of the same internal structures are called homeomorphic spaces. The concept has a precise definition that is beyond the scope of this book.

all the points within a usual circle. Even though the concept of *internal structure* is more formal than saying that the space *feels locally* like the surface of the plane, the latter is more intuitive, and so we will use it.

In Figure 6.2.1, we show parts of the surface of an almost spherical planet, like ours. All these patches on the surface of the planet *feel locally* like the surface of the plane, even though some may be flat, and some bent. Do not forget that we should imagine perceiving these surfaces from the point of view of a two-dimensional creature within them. Such a creature could not perceive the unevenness of his world. Only a three-dimensional view can reveal the way the space is curved.

FIGURE 6.2.1 A two-dimensional creature living within these surfaces, flat or rugged, would not be able to tell the difference: from within, they all feel like the surface of the plane.

Comparing a sphere and a plane, we notice that the former is *bounded* whereas the latter is not. Comparing a sphere with a space that consists of two disjoint spheres, we say that the former is *connected* whereas the latter is not. What kinds of bounded, connected two-manifolds are there? Can we list all of them?

Before we start our search we eliminate one more class of two-manifolds. Consider, for example, the space consisting of all points *strictly* within a circle. Since we are not including the circle itself, every point in this space does *feel like* a point on the plane. In this world, the circle itself may be considered as a kind of an external boundary. There is no such external boundary for the space consisting of the points on the surface of the sphere. We do not want to look for spaces with such external boundaries.* So, summarizing, we will start searching for connected, bounded, two-manifolds without external boundaries.

We have found out that any sphere, small or large, is a two-manifold of the type we want. Moreover, we noticed that bending or making any kind of unevenness on the surface of the sphere would not alter the property of being a two-manifold and it will not affect the other properties we have specified. So, there are infinitely many two-manifolds we can get by deforming the sphere without cutting or pasting. Describing all possible shapes we can get by deforming just a sphere is not feasible.

---

* The spaces without an external boundary are called *compact* spaces. Explaining the meaning of this term in more precise terms would take us too much afar, so we leave it as above. We only note that there is a slight redundancy in our requirements, since it could be shown that *compact* spaces are always bounded.

We now make a major modification in our search. Suppose that we do *not* distinguish between two homotopic two-manifolds. Then all of the two-manifolds obtained by deforming a sphere without cutting or pasting will be considered to be the same. So, we will be satisfied if we can find a list containing one representative for each class of mutually homotopic manifolds. We will call such representatives **homotopy representatives** of classes of mutually homotopic two-manifolds.

Summarizing, we want a list of homotopy representatives of all bounded, connected two-manifolds, without external boundary, and not pairwise homotopic. From now on, we will shorten the phrase "bounded, connected two-manifold, without external boundary" to the simple "two-manifold." □

Orientable Two-Manifolds

In Figures 6.2.2 to 6.2.8 we indicate how to get a complete list of all, *simple* ("*orientable*") two-manifolds (we will deal with the complicated ones in the next section). In Figure 6.2.2 we depict the familiar sphere. In the next illustration (Figure 6.2.3) we show a torus (the surface of a donut). It is easy to see that a torus fulfils all of the requirements that we have postulated: it feels like a plane from the point of view of a two-dimensional creature existing on the surface. It is bounded, it is connected, and it does not have any external boundary. So, a torus is a two-manifold of the type we consider here. Moreover, it should be *obvious* that the sphere and the torus are not homotopic: there is no way to deform the torus into a sphere without at some point pasting the hole of the torus.

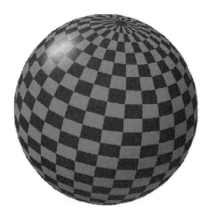

FIGURE 6.2.2 A sphere. It also represents all two-manifolds that are homotopic to the sphere.

FIGURE 6.2.3 A torus. It also represents all two-manifolds that are homotopic to the torus.

It transpires that all of the remaining representatives of the *simple* two-manifolds can be obtained from tori via an operation called **connected sum**. Here is how we get a connected sum of two tori: make circular holes on both tori, and then glue these punctured tori along the boundaries of the circular holes (see Figures 6.2.4 through 6.2.6).

The final outcome is shown in Figure 6.2.6. The surface we get is a two-manifold, and since it has two holes it is homotopic neither to a sphere nor to a torus. So, it is a new homotopy representative of two-manifolds.

FIGURE 6.2.4 Cut circular holes on two tori.

FIGURE 6.2.5 Move them closer together, ….

FIGURE 6.2.6 … and glue the surfaces along the boundaries of the holes.

The other homotopy representatives of the simple two-manifolds can be obtained by repeating the procedure of taking a connected sum. So, in the next step we take a torus and the two-manifold shown in Figure 6.2.6, cut circular holes in both of them, and glue along the boundaries of the holes. The result is shown in Figure 6.2.7, whereas the next space obtained in such a way is shown in Figure 6.2.8.

FIGURE 6.2.7 A connected sum of three tori: this is a new two-manifold.

FIGURE 6.2.8 A connected sum of four tori.

The procedure can be repeated any number of times, and we always get a brand new two-manifold that is not homotopic to any other two-manifold obtained earlier. We summarize what we have found so far in the next theorem, where we use the term *orientable* instead of the word *simple*; we will explain the terminology in the next section.

Theorem: (Classification theorem for orientable two-manifolds) Every orientable two-manifold is homotopic to a sphere, a torus, or a connected sum of (any finite number of) tori. □

The Euler Characteristic and the Genus of Two-Manifolds

We have already encountered Euler characteristic in Section 5.3: it was defined as $V - E + F$, where V is the number of vertices, E is the number of edges, and F is the number of faces on the surface of a polyhedron. We noticed there that the surfaces of all *sphere-like* polyhedra always have the Euler characteristic of exactly 2. Now we can say what it was that we meant when we said *sphere-like*: *sphere-like* means *homotopic to a sphere*.

We can easily extend the definition of the Euler characteristic to the two-manifolds we deal with in this chapter. Take, for example, a torus. Tile the torus (or any *torus-like* surface) with any kind of (somewhat bent) polygons: so the whole surface is covered with polygons and any two adjacent polygons on the surface have either a vertex or a whole edge in common. Now denote the number of vertices in this tiling by V, the number of edges by E, and the number of faces (polygons) by F. The Euler characteristic of the torus is defined to be (not surprisingly) $V - E + F$. It turns out that *it does not matter* how many polygons we use in the tiling or what kind of tiling we use on the surface of the torus: the number $V - E + F$ for a torus will *always* be equal to 0. This number is independent of not only the tiling we use but also what kind of *torus-like* surface we deal with. As long as we have a surface homotopic (deformable) to a torus, the number $V - E + F$ will always be 0.

Similarly, if we tile the surface of a sphere, or any space homotopic to a sphere, with (bent) polygons and then compute the Euler characteristic $V - E + F$, the result will always be 2. Now we see what lies behind the seeming coincidence that the Euler characteristics of the convex polyhedra we have listed in Section 5.3 are all equal to 2: the Euler characteristics of the convex polyhedra are 2, precisely because their surfaces are all homotopic to the surface of a sphere. Moreover, we see that it is the homotopy between the surfaces of the polyhedra and the surface of a sphere that matters, not the convexity of the polyhedra. Thus, for example, the surface of the stellated rhombic dodecahedron, which we have used in Section 5.5 to tile the three-dimensional space, is also of Euler characteristic equal to 2, even though it is not convex. This is true since that surface is also homotopic to the surface of a sphere.

We define the ***Euler characteristic*** of a two-manifold to be $V - E + F$, where V is the number of vertices, E is the number of edges, and F is the number of faces (polygons) in *any* tiling of that surface with any kind of polygons. As suggested by the two examples above (sphere and torus), this definition makes sense, because whatever tiling we use, the number $V - E + F$ will always be the same.

Exercise: Compute the Euler characteristic of a connected sum of two tori.

The most straightforward solution of the above exercise is to make a clay or paper model of the connected sum of two tori and then tile the surface (or draw a tiling) with polygons in any way you want. Then compute V, E, F, and finally $V - E + F$ to get the answer. There is a shorter route; for that, we need the notion of the *genus* of a two-manifold.

The *genus* of a two-manifold is the maximal number of consecutive closed circular cuts we can make on that surface without disconnecting it. For example, the genus of a sphere is 0, since any closed circular cut separates the sphere into two components (and thus, the maximal number of cuts we can make without disconnecting the sphere is 0) (Figure 6.2.9). (Note that we use the word "circular" in topological sense, so that any shape that is homotopic to a circle is *circular* in this context.)

On the other hand, the genus of any torus is 1: there is a maximum of one circular cut on the torus that will not separate it into two parts. We indicate one such torus in Figure 6.2.10. It is not hard to convince oneself that any two consecutive circular cuts would disconnect the torus.

FIGURE 6.2.9 The circular cut separates the sphere into two parts: one inside and the other outside the cut.

FIGURE 6.2.10 Cutting along that meridian will not disconnect the torus. But subsequent circular cuts will disconnect it. Consequently, the genus of the torus is 1.

There is a simple formula relating the genus and the Euler characteristic of a two-manifold. Denote the Euler characteristic of a space X by $e(X)$, and denote the genus of the same space by $g(X)$. Then we have

$$e(X) = 2 - 2g(X)$$

This formula is consistent (as it should be) with what we have discovered so far. For example, for a sphere S we have computed that $e(S) = 2$ and that $g(S) = 0$, and these two results fit the above formula. Similarly, if T stands for a torus, then $g(T) = 1$ (as seen above). Putting that in the formula yields $e(T) = 2 - 2g(T) = 2 - 2 = 0$ as claimed above.

The main property of genus and the Euler characteristic is that each of them uniquely specifies the type of the two-manifold. For example, if we have a two-manifold and if its Euler characteristic is 2, then it *has to be* that that manifold is homotopic (deformable) to a sphere.*

□

---

* The last statement is a theorem and its proof is way beyond what we do here. (It needs basic algebraic topology.)

Exercises:

1. Why is a cylindrical surface not listed in the classification theorem for two-manifolds? Why is a cylindrical surface extending without end on both sides not listed in the classification theorem for two-manifolds?

2. Imagine a connected sum of a sphere and a torus (we cut circular holes on the sphere and on the torus and then glue along the circular boundaries of the holes). Identify the type of this manifold as one mentioned in the classification theorem.

3. a. Consider the steering wheel depicted in Figure 6.2.11. This is a simple (orientable) two-manifold. To which two-manifold listed in the classification theorem is it homotopic? Visualize a homotopy. Find the genus and the Euler characteristic.

   b. Do the same for the steering wheel depicted in Figure 6.2.12.

4. Find the genus of the connected sum of two tori. Then use the formula relating it to the Euler characteristic to compute the latter.

5*. Find the genus and the Euler characteristic of the surface shown in Figure 6.2.13. We get it by drilling two circular cuts on each of the three tori and then gluing the tori along the cuts as indicated in Figure 6.2.13.

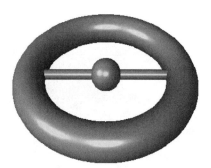

FIGURE 6.2.11 A simple steering wheel two-manifold.

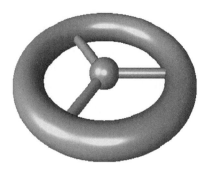

FIGURE 6.2.12 Another steering wheel two-manifold.

FIGURE 6.2.13 Three tori glued along six circular openings.

6*. In the year 1521, Magellan sailed off with a flotilla of ships, intending to circumnavigate the earth. He was killed 2 months later, but the trip was completed in 1522 by Elcano, one of the members of his crew. According to an encyclopedia, that provided a practical proof that the earth was round. Now consider the planet depicted in Figure 6.2.14.

   a. The surface (ignoring the fissures or punctures in it) of this unusual planet is a two-manifold. So, it must be homotopic to one of the two-manifolds listed in the classification theorem. Identify that two-manifold!

   b. Can you see how to get a homotopy from the two-manifold shown in Figure 6.2.14 to the two-manifold you have identified in above (6a). [*Hint:* read Example 1 in Section 6.1.]

FIGURE 6.2.14 An unusual round planet. (Based on a cartoon, *Mathematical Intelligencer* (Vol. 23, No. 1), Winter 2001. © Springer.)

6.3 NON-ORIENTABLE TWO-MANIFOLDS AND THREE-MANIFOLDS

A Torus from a Rectangle

We will now describe (in pictures) how to get a torus from a rectangular piece of flexible material. This procedure will be our main route to obtain and describe other, more complicated two-manifolds.

In Figure 6.3.1 we show a (flexible) rectangle. We will fold it and glue some pairs of edges. The two vertical edges should be glued as indicated by the arrows: the bottom of one edge with the bottom of the other, and the top of one edge with the top of the other. For example, the point A should be (eventually) glued to the point A^*.

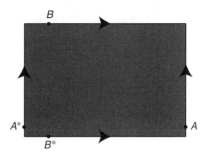

FIGURE 6.3.1 A flexible rectangle: the arrows indicate how to glue the edges.

The horizontal edges should be identified in a similar manner: each point on the top edge with the point straight below it on the bottom edge. The pair of points B and B^* is one such pair of points. Now, we slowly bend the rectangle to get into the position to perform the identification (gluing). The animation is given in the sequence of illustrations shown in Figure 6.3.2 (where we first glue the vertical edges, then we take care of the horizontal edges).

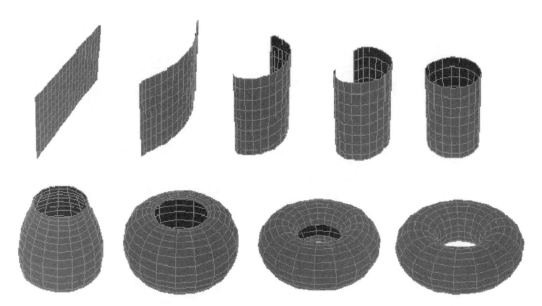

FIGURE 6.3.2 In the first row above, we describe how to identify the vertical edges of the starting rectangle, while in the next row we animate the gluing of the horizontal edges.

Möbius Band

Now, we again start with a flexible rectangular region or a rectangular piece of paper, but this time we identify only one pair of edges as indicated in Figure 6.3.3.

FIGURE 6.3.3 A flexible rectangle with two edges to be glued.

Each point A on the left edge is to be glued with the point A^* the same distance from the top as the distance from A to the bottom of the rectangle (Figure 6.3.3). The resulting space is called a **Möbius* band** and is depicted in Figure 6.3.4. Note that Möbius bands are *not* two-manifolds since neighborhoods of the points at the edge of the space do not *feel* like neighborhoods of the points in a plane.

There is one property of the Möbius band that we want to focus on: its nonorientability. As was the case a

---

* Möbius is sometimes spelled Moebius.

FIGURE 6.3.4 A Möbius band. This space has some unusual properties: for example, it has only one side. It is easy to make a model of a Möbius band from a long rectangular piece of paper by gluing a pair of its edges as explained above. You can then convince yourself that it has only one side by tracing a continuous path from a point to the same point "on the other side" along the surface of the Möbius band.

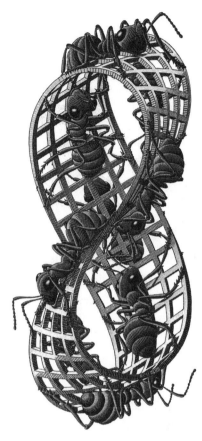

FIGURE 6.3.5 M. C. Escher. *Möbius Band II*. Woodcut, 1963.

number of times in this chapter, we will try to explain this concept in an informal manner.

Consider the graphic by Escher in Figure 6.3.5. We see a few ants on a space similar to the Möbius band. Imagine now that the space is exactly a Möbius band (as seen in Figure 6.3.4) and, moreover, imagine that the ants are two-dimensional—as if compressed completely within the two-dimensional surface of the band. Finally, suppose that the ants on the picture indicate the position of a single ant as it travels along the band. All of these requirements can be easily modeled on a Möbius band made of paper: simply sketch copies of a single ant as it moves along the surface. Make sure that in the starting sketch of your ant you somehow identify the left-hand side of the ant (say, by coloring the legs on the left-hand side with a color different from that of the legs on the right-hand side). Now watch what happens when the ant makes one circle along the surface of the Möbius band and reaches its original position *from the other side of the band*. From the perspective of the ant (completely immersed in the two-dimensional space), there are no two sides, so that upon finishing the circular journey, it will reach the very same position from which it started the journey. However, there will be one major change: what used to be ant's left-hand side has now become the right-hand side and vice versa. The orientation of the ant has changed. ☐

So, the two-dimensional creatures living on a space shaped like the Möbius band could interchange the left- and right-hand sides of their bodies by simply making a long round trip in this space. Spaces possessing this property are called ***non-orientable spaces***.

Backing up for a moment, we can now clarify the terminology we used in the preceding section: orientable, or simple two-manifolds are two-manifolds that do not possess orientation-reversing paths. All of the manifolds we encountered in Section 6.2 satisfy this property.

Klein Bottle

As we will see in a moment, some two-manifolds are non-orientable. We will now provide a complete list of all non-orientable (connected, bounded, without external boundaries) two-manifolds. We start with arguably, the simplest one: the ***Klein bottle***. We can get a Klein bottle by again gluing pairs of edges of a flexible filled rectangle. The recipe is given in Figure 6.3.6 and in its caption.

The outcome, a Klein bottle, is shown in Figure 6.3.7 and Figure 6.3.8.

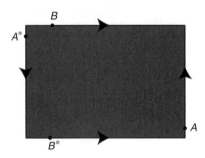

FIGURE 6.3.6 The horizontal edges are identified in the usual way (identifying only these edges will give us an open cylinder). The vertical edges are to be identified the same way as for the Möbius band: in the opposite direction.

FIGURE 6.3.7 A Klein bottle.

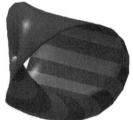

FIGURE 6.3.8 One more Klein bottle.

It appears that there is some self-intersection in the Klein bottle. This is not a characteristic of the internal structure of the Klein bottle. Rather, it is due to lack of space for visualization. The Klein bottle cannot be depicted properly in three dimensions. We need four dimensions to show it without any self-intersection. Unfortunately, four dimensions are beyond our visual perception. The problem of explaining this phenomenon is similar to the problem of explaining to a two-dimensional creature that a two-dimensional representation of a knot shows self-intersections, where no self-intersections exist in the proper, three-dimensional home of the knot.

Every Klein bottle is a two-manifold. However, unlike the two-manifolds described in the previous section, this one is non-orientable. In Figure 6.3.9, we show an artistic rendering of a Klein bottle.

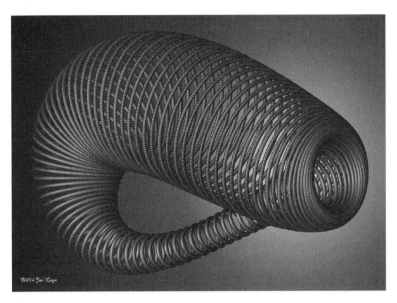

FIGURE 6.3.9 **(See color insert following page 144.)** Jos Leys. *Klein Bottle*, 2004. □

Projective Plane

We will now describe another important non-orientable two-manifold. As was the case with the previous two-manifolds, this one can (in principle) also be obtained from a flexible rectangle by identifying the edges.* The exact recipe is given in Figure 6.3.10 and in the explanation given below.

Notice that we glue the pair of vertical edges in opposite directions (so that the point A is identified with A^*) and that we do the same with the pair of horizontal edges (the point B is identified with the point B^*). The resulting space is called the ***projective plane***. As was the case with the Klein bottle, this is also a two-manifold, which has its home in four dimensions: our two-dimensional rendering of the

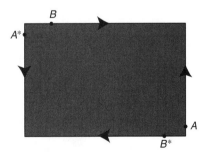

FIGURE 6.3.10 A projective plane from a filled rectangle.

---

* In fact, every two-manifold can be obtained from a (very flexible) polygonal piece with evenly many edges, by gluing some of the pairs of edges.

projective plane (Figure 6.3.11) shows some self-intersections that do not exist in its proper setting. In Figures 6.3.11 and 6.3.12, we depict a projective plane; the wire-frame version of the same projective plane shown in Figure 6.3.12 shows more clearly the intricacy and the symmetry of the structure.

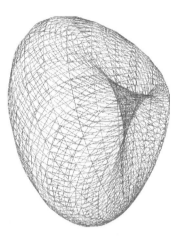

FIGURE 6.3.11 A projective plane: the lack of space (in three dimensions) is now more acute, and we see significant self-intersection where no self-intersection exists.

FIGURE 6.3.12 A wire-frame version of the same projective plane: we see how symmetric this object is.

Classification of Non-orientable Two-Manifolds

We know from the previous section how to get new, orientable two-manifolds from old orientable two-manifolds through their connected sums. Recall that we get a connected sum by cutting two circular openings on the two surfaces and then gluing along the circles bounding the two cuts. Precisely the same operation can be performed to non-orientable two-manifolds. For example, cutting out circular openings in two projective planes and then gluing along the circles bounding the cuts produces a connected sum of two projective planes. In Figure 6.3.13, we show a connected sum of two projective planes.

We can continue this procedure and attach another projective plane to the space depicted in Figure 6.3.13 in the same manner: that would give us a connected sum of

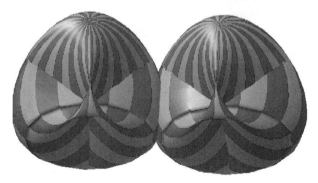

FIGURE 6.3.13 A connected sum of two projective planes.

three projective planes. Repeating once again yields a connected sum of four projective planes. We can go on in this manner to get connected sums of any number of projective planes. It turns out that these are the only homotopy representatives of the non-orientable two-manifolds. We summarize this in the next theorem (in which, as we have indicted earlier, we use the term *two-manifold* as short for "connected, bounded two-manifold without external boundary").

> **Theorem: (Classification theorem for non-orientable two-manifolds)** Every non-orientable two-manifold is homotopic to a projective plane or to a connected sum (some finite number) of projective planes.

Most of the proof of the two classification theorems (Sections 6.2 and 6.3) was done during the early years of the twentieth century. The last patch of the proof was fixed by Rado in 1925.

Have we missed the Klein bottle in this theorem? No: it can be shown that a connected sum of two projective planes (Figure 6.3.13) is homotopic (deformable) to a Klein bottle (Figures 6.3.7 through 6.3.8). Figure 6.3.14 shows another Jos Leys' artwork depicting a two-manifold (assuming smooth surface where rings are shown). This two-manifold is homotopic (within four dimensions) to a torus (Exercise 7). □

The definition of the Euler characteristic for non-orientable two-manifolds is the same as the one given for orientable two-manifolds (see Exercise 4, Section 6.3). However, if we extend the loose definition of the genus of an orientable surface to non-orientable manifolds without any changes (the maximal number of closed cuts that can be performed successively without disconnecting the surface) then we get some unexpected results. For example, under that definition, the genus of a Klein bottle is infinity!

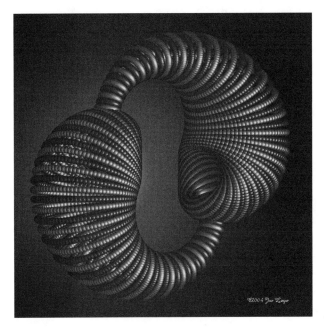

FIGURE 6.3.14 **(See color insert following page 144.)** Jos Leys. *Bonan-Jeener's Klein Surface 1*, 2004.

A Brief Overture into the Realm of Three-manifolds

Start with a disk (the circle, together with its interior points) and then glue each point in the upper half-circle with the point in the lower half-circle straight below it (see Figure 6.3.15). The effect is the same as zipping the disk shut with a zipper along its boundary, and the result is a two-manifold that is homotopic to a sphere (see Figure 6.3.16).

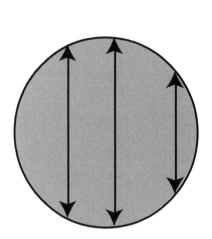

FIGURE 6.3.15 The arrows indicate how to glue the top and the bottom points.

FIGURE 6.3.16 The effect is the same as zipping the disk shut (the zipper, of course, is just to illustrate that point). We get a sphere.

As we see, starting with a space with a boundary in two dimensions (the disk), then gluing along that boundary, produced a two-manifold. (We will now mimic that procedure one dimension higher.) Start with a sphere together with its interior points (a ball) and glue each point in the upper hemisphere with the point in the lower hemisphere that is on the same vertical line (see Figure 6.3.17).

The resulting space is not easy to visualize anymore, since it cannot be put in three dimensions (we need four dimensions to see it). However, it is not very hard to convince ourselves (we will do that in the next paragraph) that the resulting space is a ***three-manifold***, that is, it feels at every point the same as the *usual* three-dimensional space that we perceive. Precisely because we experience a three-manifold through our basic perceptions, describing the possible shapes of bounded, connected three-manifolds (similar definitions as

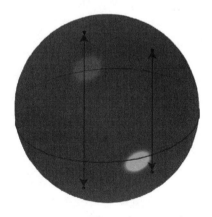

FIGURE 6.3.17 Glue the top hemisphere to the bottom hemisphere (the equator points are not affected). The result is a three-manifold.

for two-manifolds) is an important prob-lem. Some major progress in the direction of solving this problem was reported lately (early twenty-first century).

Let us try to understand why the space obtained by gluing pairs of points on the surface of a ball as described above is indeed a three-manifold. To start with, the close neighborhood of each point in interior of the sphere *feels* the same as the neigh-borhood of the points in the *usual* three-dimensional space. Now we pay attention to the points on the sphere. Gluing the points on the sphere as indicated has the same effect as gluing along the bounding planes of the two half-spaces as shown in Figure 6.3.18. One of these two spaces consists of all points in the *usual* three-dimensional space above and on a given plane, and the second space is made of all

FIGURE 6.3.18 We see two parallel planes in perspective. The first space consists of all points below or on the bottom plane, while the second space is made of the points above or on the top plane. When we glue the planes, the result is the usual three-dimensional space.

points below and on another plane. So, after the gluing is performed, the points on the bounding planes merge into points that *feel* the same as any other point in the three space, the top half of their neighborhood supplied by the first space, whereas the bottom half comes from the second space. The same effect happens in the case of gluing the upper and the lower hemispheres of a ball in the manner described above.

It can be shown that the space we get after gluing the top hemisphere with the bottom hemisphere of a ball (as indicated in Figure 6.3.17) is homotopic to the three-dimensional sphere, consisting of all points in four dimensions that are equidistant from a fixed point (the center of the three-dimensional sphere). We note that this sphere is called three-dimensional not because it exists in three dimensions, but because it *feels* locally like the usual three-dimensional space (in the same way that the usual sphere *feels* locally like a plane). So, it is a three-manifold. We see that a three-manifold, a space that *feels* like our *usual* three-dimensional space, could be bounded.

Two more examples of three-manifolds are described below. The first one is obtained by gluing the opposite faces in a cube. In Figure 6.3.19, we show one way of gluing the top and the bottom faces of a cube (the other pairs of opposite faces should also be glued). We again encounter the problem of *lack of space*, and so it is not easy to visualize. We only show (Figure 6.3.20) what happens when we glue the top and the bottom faces as indicated (with the other two pairs of opposite faces still to be glued).

What we get (after gluing *all* the pairs of the opposite faces in the cube—not shown in Figures 6.3.19 and 6.3.20) is a brand new bounded compact three-manifold. This being complicated enough, it is even more complicated if the identification of the opposite faces is done in *opposite* face-orientation (see Figure 6.3.21).

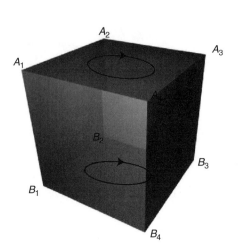

FIGURE 6.3.19 After gluing the top face with the bottom face, the vertices A_1, A_2, A_3, and A_4 are identified with the vertices B_1, B_2, B_3, and B_4, respectively.

FIGURE 6.3.20 When we glue the top and the bottom faces, we get a filled torus-like object. From then on we glue (what used to be) the other two pairs of opposite faces. Properly visualizing the resulting space is not possible in three dimensions.

We note again that Figures 6.3.21, 6.3.22 and 6.3.23 depict only a part of the gluing process: we still need to glue the other two pairs of opposite faces. In the end, we do get a three-manifold. This three-manifold is non-orientable. A three-manifold is non-orientable if there is an orientation-reversing circular trajectory in that space, meaning that if we take a trip along that trajectory, we would end up at the starting position with our left hand becoming right and vice versa. Going straight up from the middle of the square and then

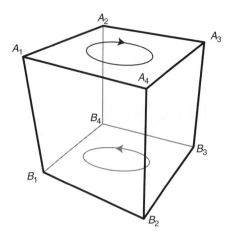

FIGURE 6.3.21 Gluing the top face with the bottom face in the opposite orientation, takes the vertices A_1, A_2, A_3, and A_4 to the vertices B_1, B_2, B_3, and B_4, respectively.

FIGURE 6.3.22 We pull the top face of the cube and stretch the cube as shown here.

FIGURE 6.3.23 Even visualizing the space we get after we identify only one pair of opposite faces in this manner cannot be done properly in three dimensions. We get a space that looks like a *filled* Klein bottle. A proper visualization does not have any self-intersections.

ending again at the same position (keeping in mind that the opposite faces are identified as indicated) is such an orientation-reversing circular trip in the manifold partially described in Figures 6.3.21, 6.3.22, and 6.3.23. It is conceivable that *our* three-dimensional space (the space in which we exist materially) is non-orientable. □

The two sculptures in Figures 6.3.24 and 6.3.25 are clearly related to the subject of three-manifolds.

FIGURE 6.3.24 John Robinson. *Dependent beings*, 1980, polished bronze; displayed in the Centre de Recerca, Institut D'Estudis Catalans, Barcelona, and in the Philip Trust Collection, U.K.

FIGURE 6.3.25 Charles O. Perry. *Zero*, Norwalk, CT, bronze, 26 in. diameter, 2003.

Exercises:

1. Show that every Möbius band is homotopic to a circle.

2. Show that the genus of a Möbius band is infinity. That is, show that we can make arbitrarily many consecutive closed circular cuts without disconnecting the Möbius band. [*Hint*: take a long rectangular piece of paper, glue the pair of short edges according to the recipe given in Figure 6.3.3, and then take scissors and start cutting!]

3. Show that the genus of both the Klein bottle and the projective plane are infinity. That is, show that one can make arbitrarily many circular cuts on these two-manifolds without disconnecting them. [*Hint*: it is easier to see what happens if you do the cut on the rectangles shown in Figures 6.3.3 (for the Klein bottle) and 6.3.6 (for the projective plane), and then identify the edges.]

4. In Figure 6.3.26 we see a tiled rectangle. Pairs of edges of the rectangle are to be glued as indicated, resulting in a Klein bottle. The tiling of the rectangle with the triangles as shown does not transfer to a tiling of the Klein bottle after gluing the edges! Why not? [*Hint*: in a tiling, adjacent polygons share a common edge or a common vertex. Find two triangles in the Klein bottle that share two vertices and no common edge.]

5. In Figure 6.3.27, we show a tiling of a rectangle that does transfer to a tiling of the Klein bottle we get by identifying the edges of the rectangle as indicated. Carefully count the number of polygons, the number of edges, and the number of vertices in this tiling of the Klein bottle. Find the Euler characteristic of the Klein bottle. (Be careful when counting the edges and the vertices! Some of the edges and some of the vertices in the tiling of the rectangle will be glued in the Klein bottle!)

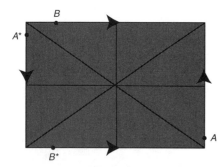

FIGURE 6.3.26 The tiling of this rectangle does not become a tiling of the Klein bottle.

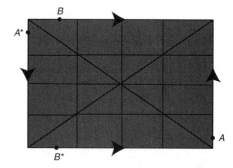

FIGURE 6.3.27 This tiling of the rectangle will become a tiling of the Klein bottle.

6. Take a ball (a sphere together with its interior points) and glue (in your mind; do not try to visualize!) the two points in each pair of antipodal points on the surface of the ball.
   a. Convince yourself that the resulting space is a three-manifold (i.e., it *feels* like the three-dimensional space at every point). This space is called the *projective space*.
   b. Is the projective space orientable or non-orientable? In case it is non-orientable, find one orientation-reversing trip.

7*. Consider the object made of two Klein bottles as depicted in Jos Leys' artwork (Figure 6.3.14). Explain why it is homotopic within four dimensions to a torus. [*Hint*: this requires an argument along the lines of the preliminaries in Example 2, Section 6.1. See also Exercise 7, Section 6.1.]

Index

A

Addition
 complex numbers, 102
 matrices, 46
 of points in plane, 43
Ammann tilings, 80
Angle
 duplication of, 8
 hyperbolic construction, 161–162
 of displacement, 27
 of rotation, 34
 positive/negative, 36–37
Antiprisms, 201
Antiquity, problems of, 13
Aperiodic tiling, 79–80. *See also* Tilings
Apex, 208
Archimedean solid, 201–202
 space tilings, 219
Archimedean tilings, 78–79
Argument, of a complex number, 104
Athenian school of mathematics, 4
Avatamsaka Sutra, 150–151
Axes
 Cartesian coordinates, 42
 cones, 208
Axioms
 defined, 4
 Euclidean geometry, 5–6
 Hyperbolic geometry, 144

B

Ball, 206
Basic translations, wallpaper designs,
 72–73
Beckhampton's crop circles, 66–67
Bijection, defined, 33
Binary system, 134
Binet formula, 29
Bounded, sets of points in plane, 124

Bounded space, 231
Brunelleschi, Filippo, 175

C

Cantor set, 121–122
Cardioid, construction of, 214–215
Cartesian coordinate system, *see* Descartes
 coordinate system, 42
Cellular automata
 Game of Life, 137
 notation, 134
 one-dimensional, 133
 Rule 18, 131–133
 Rule 86, 133–134
 Rule X, 135–137
 three-dimensional, 220
 two-dimensional, 134
Centerline, 160
Center of rotation, 34
Central similarities
 center of, 92
 defined, 92
Chirals, 201
Circles
 center of partially given circle, construction of, 9
 perspective drawing, 193
 sketching using tangents, 210–211
 touching another circle, construction of, 9–11
Circle reflections, *See* Inversions
Circular inversions, *See* Inversions
Classification theorem
 frieze patterns, 66–72
 non-orientable two-manifolds, 242–243
 orientable two-manifolds, 233
 plane symmetries, 37–38
 similarities, 95–96
Class of parallel lines, perspective drawing, 172–173
Colors, Julia sets and, 127–128
Common sense axioms, Euclidean geometry, 5